Volcanoes: A Very Short Introduction

VERY SHORT INTRODUCTIONS are for anyone wanting a stimulating and accessible way into a new subject. They are written by experts, and have been translated into more than 45 different languages.

The series began in 1995, and now covers a wide variety of topics in every discipline. The VSI library currently contains over 650 volumes—a Very Short Introduction to everything from Psychology and Philosophy of Science to American History and Relativity—and continues to grow in every subject area.

Very Short Introductions available now:

ABOLITIONISM Richard S. Newman
THE ABRAHAMIC RELIGIONS Charles L. Cohen
ACCOUNTING Christopher Nobes
ADAM SMITH Christopher J. Berry
ADOLESCENCE Peter K. Smith
ADVERTISING Winston Fletcher
AERIAL WARFARE Frank Ledwidge
AESTHETICS Bence Nanay
AFRICAN AMERICAN RELIGION Eddie S. Glaude Jr
AFRICAN HISTORY John Parker and Richard Rathbone
AFRICAN POLITICS Ian Taylor
AFRICAN RELIGIONS Jacob K. Olupona
AGEING Nancy A. Pachana
AGNOSTICISM Robin Le Poidevin
AGRICULTURE Paul Brassley and Richard Soffe
ALBERT CAMUS Oliver Gloag
ALEXANDER THE GREAT Hugh Bowden
ALGEBRA Peter M. Higgins
AMERICAN BUSINESS HISTORY Walter A. Friedman
AMERICAN CULTURAL HISTORY Eric Avila
AMERICAN FOREIGN RELATIONS Andrew Preston
AMERICAN HISTORY Paul S. Boyer
AMERICAN IMMIGRATION David A. Gerber
AMERICAN LEGAL HISTORY G. Edward White
AMERICAN MILITARY HISTORY Joseph T. Glatthaar
AMERICAN NAVAL HISTORY Craig L. Symonds
AMERICAN POLITICAL HISTORY Donald Critchlow
AMERICAN POLITICAL PARTIES AND ELECTIONS L. Sandy Maisel
AMERICAN POLITICS Richard M. Valelly
THE AMERICAN PRESIDENCY Charles O. Jones
THE AMERICAN REVOLUTION Robert J. Allison
AMERICAN SLAVERY Heather Andrea Williams
THE AMERICAN WEST Stephen Aron
AMERICAN WOMEN'S HISTORY Susan Ware
ANAESTHESIA Aidan O'Donnell
ANALYTIC PHILOSOPHY Michael Beaney
ANARCHISM Colin Ward
ANCIENT ASSYRIA Karen Radner
ANCIENT EGYPT Ian Shaw
ANCIENT EGYPTIAN ART AND ARCHITECTURE Christina Riggs
ANCIENT GREECE Paul Cartledge
THE ANCIENT NEAR EAST Amanda H. Podany
ANCIENT PHILOSOPHY Julia Annas
ANCIENT WARFARE Harry Sidebottom
ANGELS David Albert Jones
ANGLICANISM Mark Chapman
THE ANGLO-SAXON AGE John Blair

ANIMAL BEHAVIOUR Tristram D. Wyatt
THE ANIMAL KINGDOM Peter Holland
ANIMAL RIGHTS David DeGrazia
THE ANTARCTIC Klaus Dodds
ANTHROPOCENE Erle C. Ellis
ANTISEMITISM Steven Beller
ANXIETY Daniel Freeman and Jason Freeman
THE APOCRYPHAL GOSPELS Paul Foster
APPLIED MATHEMATICS Alain Goriely
ARCHAEOLOGY Paul Bahn
ARCHITECTURE Andrew Ballantyne
ARISTOCRACY William Doyle
ARISTOTLE Jonathan Barnes
ART HISTORY Dana Arnold
ART THEORY Cynthia Freeland
ARTIFICIAL INTELLIGENCE Margaret A. Boden
ASIAN AMERICAN HISTORY Madeline Y. Hsu
ASTROBIOLOGY David C. Catling
ASTROPHYSICS James Binney
ATHEISM Julian Baggini
THE ATMOSPHERE Paul I. Palmer
AUGUSTINE Henry Chadwick
AUSTRALIA Kenneth Morgan
AUTISM Uta Frith
AUTOBIOGRAPHY Laura Marcus
THE AVANT GARDE David Cottington
THE AZTECS Davíd Carrasco
BABYLONIA Trevor Bryce
BACTERIA Sebastian G. B. Amyes
BANKING John Goddard and John O. S. Wilson
BARTHES Jonathan Culler
THE BEATS David Sterritt
BEAUTY Roger Scruton
BEHAVIOURAL ECONOMICS Michelle Baddeley
BESTSELLERS John Sutherland
THE BIBLE John Riches
BIBLICAL ARCHAEOLOGY Eric H. Cline
BIG DATA Dawn E. Holmes
BIOGEOGRAPHY Mark V. Lomolino
BIOGRAPHY Hermione Lee
BIOMETRICS Michael Fairhurst
BLACK HOLES Katherine Blundell
BLOOD Chris Cooper
THE BLUES Elijah Wald
THE BODY Chris Shilling
THE BOOK OF COMMON PRAYER Brian Cummings
THE BOOK OF MORMON Terryl Givens
BORDERS Alexander C. Diener and Joshua Hagen
THE BRAIN Michael O'Shea
BRANDING Robert Jones
THE BRICS Andrew F. Cooper
THE BRITISH CONSTITUTION Martin Loughlin
THE BRITISH EMPIRE Ashley Jackson
BRITISH POLITICS Tony Wright
BUDDHA Michael Carrithers
BUDDHISM Damien Keown
BUDDHIST ETHICS Damien Keown
BYZANTIUM Peter Sarris
C. S. LEWIS James Como
CALVINISM Jon Balserak
CANADA Donald Wright
CANCER Nicholas James
CAPITALISM James Fulcher
CATHOLICISM Gerald O'Collins
CAUSATION Stephen Mumford and Rani Lill Anjum
THE CELL Terence Allen and Graham Cowling
THE CELTS Barry Cunliffe
CHAOS Leonard Smith
CHARLES DICKENS Jenny Hartley
CHEMISTRY Peter Atkins
CHILD PSYCHOLOGY Usha Goswami
CHILDREN'S LITERATURE Kimberley Reynolds
CHINESE LITERATURE Sabina Knight
CHOICE THEORY Michael Allingham
CHRISTIAN ART Beth Williamson
CHRISTIAN ETHICS D. Stephen Long
CHRISTIANITY Linda Woodhead
CIRCADIAN RHYTHMS Russell Foster and Leon Kreitzman
CITIZENSHIP Richard Bellamy
CIVIL ENGINEERING David Muir Wood
CLASSICAL LITERATURE William Allan
CLASSICAL MYTHOLOGY Helen Morales

CLASSICS Mary Beard and John Henderson
CLAUSEWITZ Michael Howard
CLIMATE Mark Maslin
CLIMATE CHANGE Mark Maslin
CLINICAL PSYCHOLOGY Susan Llewelyn and Katie Aafjes-van Doorn
COGNITIVE NEUROSCIENCE Richard Passingham
THE COLD WAR Robert McMahon
COLONIAL AMERICA Alan Taylor
COLONIAL LATIN AMERICAN LITERATURE Rolena Adorno
COMBINATORICS Robin Wilson
COMEDY Matthew Bevis
COMMUNISM Leslie Holmes
COMPARATIVE LITERATURE Ben Hutchinson
COMPLEXITY John H. Holland
THE COMPUTER Darrel Ince
COMPUTER SCIENCE Subrata Dasgupta
CONCENTRATION CAMPS Dan Stone
CONFUCIANISM Daniel K. Gardner
THE CONQUISTADORS Matthew Restall and Felipe Fernández-Armesto
CONSCIENCE Paul Strohm
CONSCIOUSNESS Susan Blackmore
CONTEMPORARY ART Julian Stallabrass
CONTEMPORARY FICTION Robert Eaglestone
CONTINENTAL PHILOSOPHY Simon Critchley
COPERNICUS Owen Gingerich
CORAL REEFS Charles Sheppard
CORPORATE SOCIAL RESPONSIBILITY Jeremy Moon
CORRUPTION Leslie Holmes
COSMOLOGY Peter Coles
COUNTRY MUSIC Richard Carlin
CRIME FICTION Richard Bradford
CRIMINAL JUSTICE Julian V. Roberts
CRIMINOLOGY Tim Newburn
CRITICAL THEORY Stephen Eric Bronner
THE CRUSADES Christopher Tyerman
CRYPTOGRAPHY Fred Piper and Sean Murphy
CRYSTALLOGRAPHY A. M. Glazer
THE CULTURAL REVOLUTION Richard Curt Kraus
DADA AND SURREALISM David Hopkins
DANTE Peter Hainsworth and David Robey
DARWIN Jonathan Howard
THE DEAD SEA SCROLLS Timothy H. Lim
DECADENCE David Weir
DECOLONIZATION Dane Kennedy
DEMENTIA Kathleen Taylor
DEMOCRACY Bernard Crick
DEMOGRAPHY Sarah Harper
DEPRESSION Jan Scott and Mary Jane Tacchi
DERRIDA Simon Glendinning
DESCARTES Tom Sorell
DESERTS Nick Middleton
DESIGN John Heskett
DEVELOPMENT Ian Goldin
DEVELOPMENTAL BIOLOGY Lewis Wolpert
THE DEVIL Darren Oldridge
DIASPORA Kevin Kenny
DICTIONARIES Lynda Mugglestone
DINOSAURS David Norman
DIPLOMACY Joseph M. Siracusa
DOCUMENTARY FILM Patricia Aufderheide
DREAMING J. Allan Hobson
DRUGS Les Iversen
DRUIDS Barry Cunliffe
DYNASTY Jeroen Duindam
DYSLEXIA Margaret J. Snowling
EARLY MUSIC Thomas Forrest Kelly
THE EARTH Martin Redfern
EARTH SYSTEM SCIENCE Tim Lenton
ECOLOGY Jaboury Ghazoul
ECONOMICS Partha Dasgupta
EDUCATION Gary Thomas
EGYPTIAN MYTH Geraldine Pinch
EIGHTEENTH-CENTURY BRITAIN Paul Langford
THE ELEMENTS Philip Ball
ÉMILE ZOLA Brian Nelson
EMOTION Dylan Evans
EMPIRE Stephen Howe
ENERGY SYSTEMS Nick Jenkins
ENGELS Terrell Carver

ENGINEERING David Blockley
THE ENGLISH LANGUAGE Simon Horobin
ENGLISH LITERATURE Jonathan Bate
THE ENLIGHTENMENT John Robertson
ENTREPRENEURSHIP Paul Westhead and Mike Wright
ENVIRONMENTAL ECONOMICS Stephen Smith
ENVIRONMENTAL ETHICS Robin Attfield
ENVIRONMENTAL LAW Elizabeth Fisher
ENVIRONMENTAL POLITICS Andrew Dobson
ENZYMES Paul Engel
EPICUREANISM Catherine Wilson
EPIDEMIOLOGY Rodolfo Saracci
ETHICS Simon Blackburn
ETHNOMUSICOLOGY Timothy Rice
THE ETRUSCANS Christopher Smith
EUGENICS Philippa Levine
THE EUROPEAN UNION Simon Usherwood and John Pinder
EUROPEAN UNION LAW Anthony Arnull
EVOLUTION Brian and Deborah Charlesworth
EXISTENTIALISM Thomas Flynn
EXPLORATION Stewart A. Weaver
EXTINCTION Paul B. Wignall
THE EYE Michael Land
FAIRY TALE Marina Warner
FAMILY LAW Jonathan Herring
FASCISM Kevin Passmore
FASHION Rebecca Arnold
FEDERALISM Mark J. Rozell and Clyde Wilcox
FEMINISM Margaret Walters
FILM Michael Wood
FILM MUSIC Kathryn Kalinak
FILM NOIR James Naremore
FIRE Andrew C. Scott
THE FIRST WORLD WAR Michael Howard
FOLK MUSIC Mark Slobin
FOOD John Krebs
FORENSIC PSYCHOLOGY David Canter
FORENSIC SCIENCE Jim Fraser
FORESTS Jaboury Ghazoul
FOSSILS Keith Thomson
FOUCAULT Gary Gutting
THE FOUNDING FATHERS R. B. Bernstein
FRACTALS Kenneth Falconer
FREE SPEECH Nigel Warburton
FREE WILL Thomas Pink
FREEMASONRY Andreas Önnerfors
FRENCH LITERATURE John D. Lyons
FRENCH PHILOSOPHY Stephen Gaukroger and Knox Peden
THE FRENCH REVOLUTION William Doyle
FREUD Anthony Storr
FUNDAMENTALISM Malise Ruthven
FUNGI Nicholas P. Money
THE FUTURE Jennifer M. Gidley
GALAXIES John Gribbin
GALILEO Stillman Drake
GAME THEORY Ken Binmore
GANDHI Bhikhu Parekh
GARDEN HISTORY Gordon Campbell
GENES Jonathan Slack
GENIUS Andrew Robinson
GENOMICS John Archibald
GEOFFREY CHAUCER David Wallace
GEOGRAPHY John Matthews and David Herbert
GEOLOGY Jan Zalasiewicz
GEOPHYSICS William Lowrie
GEOPOLITICS Klaus Dodds
GEORGE BERNARD SHAW Christopher Wixson
GERMAN LITERATURE Nicholas Boyle
GERMAN PHILOSOPHY Andrew Bowie
THE GHETTO Bryan Cheyette
GLACIATION David J. A. Evans
GLOBAL CATASTROPHES Bill McGuire
GLOBAL ECONOMIC HISTORY Robert C. Allen
GLOBAL ISLAM Nile Green
GLOBALIZATION Manfred Steger
GOD John Bowker
GOETHE Ritchie Robertson
THE GOTHIC Nick Groom
GOVERNANCE Mark Bevir
GRAVITY Timothy Clifton
THE GREAT DEPRESSION AND THE NEW DEAL Eric Rauchway

HABERMAS James Gordon Finlayson
THE HABSBURG EMPIRE
Martyn Rady
HAPPINESS Daniel M. Haybron
THE HARLEM RENAISSANCE
Cheryl A. Wall
THE HEBREW BIBLE AS LITERATURE
Tod Linafelt
HEGEL Peter Singer
HEIDEGGER Michael Inwood
THE HELLENISTIC AGE
Peter Thonemann
HEREDITY John Waller
HERMENEUTICS Jens Zimmermann
HERODOTUS Jennifer T. Roberts
HIEROGLYPHS Penelope Wilson
HINDUISM Kim Knott
HISTORY John H. Arnold
THE HISTORY OF ASTRONOMY
Michael Hoskin
THE HISTORY OF CHEMISTRY
William H. Brock
THE HISTORY OF CHILDHOOD
James Marten
THE HISTORY OF CINEMA
Geoffrey Nowell-Smith
THE HISTORY OF LIFE Michael Benton
THE HISTORY OF MATHEMATICS
Jacqueline Stedall
THE HISTORY OF MEDICINE
William Bynum
THE HISTORY OF PHYSICS
J. L. Heilbron
THE HISTORY OF TIME
Leofranc Holford-Strevens
HIV AND AIDS Alan Whiteside
HOBBES Richard Tuck
HOLLYWOOD Peter Decherney
THE HOLY ROMAN EMPIRE
Joachim Whaley
HOME Michael Allen Fox
HOMER Barbara Graziosi
HORMONES Martin Luck
HUMAN ANATOMY
Leslie Klenerman
HUMAN EVOLUTION Bernard Wood
HUMAN RIGHTS Andrew Clapham
HUMANISM Stephen Law
HUME A. J. Ayer
HUMOUR Noël Carroll
THE ICE AGE Jamie Woodward
IDENTITY Florian Coulmas
IDEOLOGY Michael Freeden
THE IMMUNE SYSTEM
Paul Klenerman
INDIAN CINEMA
Ashish Rajadhyaksha
INDIAN PHILOSOPHY Sue Hamilton
THE INDUSTRIAL REVOLUTION
Robert C. Allen
INFECTIOUS DISEASE Marta L. Wayne
and Benjamin M. Bolker
INFINITY Ian Stewart
INFORMATION Luciano Floridi
INNOVATION Mark Dodgson and
David Gann
INTELLECTUAL PROPERTY
Siva Vaidhyanathan
INTELLIGENCE Ian J. Deary
INTERNATIONAL LAW Vaughan Lowe
INTERNATIONAL MIGRATION
Khalid Koser
INTERNATIONAL RELATIONS
Christian Reus-Smit
INTERNATIONAL SECURITY
Christopher S. Browning
IRAN Ali M. Ansari
ISLAM Malise Ruthven
ISLAMIC HISTORY Adam Silverstein
ISOTOPES Rob Ellam
ITALIAN LITERATURE
Peter Hainsworth and David Robey
JESUS Richard Bauckham
JEWISH HISTORY David N. Myers
JOURNALISM Ian Hargreaves
JUDAISM Norman Solomon
JUNG Anthony Stevens
KABBALAH Joseph Dan
KAFKA Ritchie Robertson
KANT Roger Scruton
KEYNES Robert Skidelsky
KIERKEGAARD Patrick Gardiner
KNOWLEDGE Jennifer Nagel
THE KORAN Michael Cook
KOREA Michael J. Seth
LAKES Warwick F. Vincent
LANDSCAPE ARCHITECTURE
Ian H. Thompson
LANDSCAPES AND
GEOMORPHOLOGY
Andrew Goudie and Heather Viles
LANGUAGES Stephen R. Anderson

LATE ANTIQUITY Gillian Clark
LAW Raymond Wacks
THE LAWS OF THERMODYNAMICS Peter Atkins
LEADERSHIP Keith Grint
LEARNING Mark Haselgrove
LEIBNIZ Maria Rosa Antognazza
LEO TOLSTOY Liza Knapp
LIBERALISM Michael Freeden
LIGHT Ian Walmsley
LINCOLN Allen C. Guelzo
LINGUISTICS Peter Matthews
LITERARY THEORY Jonathan Culler
LOCKE John Dunn
LOGIC Graham Priest
LOVE Ronald de Sousa
MACHIAVELLI Quentin Skinner
MADNESS Andrew Scull
MAGIC Owen Davies
MAGNA CARTA Nicholas Vincent
MAGNETISM Stephen Blundell
MALTHUS Donald Winch
MAMMALS T. S. Kemp
MANAGEMENT John Hendry
MAO Delia Davin
MARINE BIOLOGY Philip V. Mladenov
THE MARQUIS DE SADE John Phillips
MARTIN LUTHER Scott H. Hendrix
MARTYRDOM Jolyon Mitchell
MARX Peter Singer
MATERIALS Christopher Hall
MATHEMATICAL FINANCE Mark H. A. Davis
MATHEMATICS Timothy Gowers
MATTER Geoff Cottrell
THE MAYA Matthew Restall and Amara Solari
THE MEANING OF LIFE Terry Eagleton
MEASUREMENT David Hand
MEDICAL ETHICS Michael Dunn and Tony Hope
MEDICAL LAW Charles Foster
MEDIEVAL BRITAIN John Gillingham and Ralph A. Griffiths
MEDIEVAL LITERATURE Elaine Treharne
MEDIEVAL PHILOSOPHY John Marenbon
MEMORY Jonathan K. Foster
METAPHYSICS Stephen Mumford
METHODISM William J. Abraham
THE MEXICAN REVOLUTION Alan Knight
MICHAEL FARADAY Frank A. J. L. James
MICROBIOLOGY Nicholas P. Money
MICROECONOMICS Avinash Dixit
MICROSCOPY Terence Allen
THE MIDDLE AGES Miri Rubin
MILITARY JUSTICE Eugene R. Fidell
MILITARY STRATEGY Antulio J. Echevarria II
MINERALS David Vaughan
MIRACLES Yujin Nagasawa
MODERN ARCHITECTURE Adam Sharr
MODERN ART David Cottington
MODERN CHINA Rana Mitter
MODERN DRAMA Kirsten E. Shepherd-Barr
MODERN FRANCE Vanessa R. Schwartz
MODERN INDIA Craig Jeffrey
MODERN IRELAND Senia Pašeta
MODERN ITALY Anna Cento Bull
MODERN JAPAN Christopher Goto-Jones
MODERN LATIN AMERICAN LITERATURE Roberto González Echevarría
MODERN WAR Richard English
MODERNISM Christopher Butler
MOLECULAR BIOLOGY Aysha Divan and Janice A. Royds
MOLECULES Philip Ball
MONASTICISM Stephen J. Davis
THE MONGOLS Morris Rossabi
MONTAIGNE William M. Hamlin
MOONS David A. Rothery
MORMONISM Richard Lyman Bushman
MOUNTAINS Martin F. Price
MUHAMMAD Jonathan A. C. Brown
MULTICULTURALISM Ali Rattansi
MULTILINGUALISM John C. Maher
MUSIC Nicholas Cook
MYTH Robert A. Segal
NAPOLEON David Bell
THE NAPOLEONIC WARS Mike Rapport
NATIONALISM Steven Grosby
NATIVE AMERICAN LITERATURE Sean Teuton

NAVIGATION Jim Bennett
NAZI GERMANY Jane Caplan
NELSON MANDELA Elleke Boehmer
NEOLIBERALISM Manfred Steger and Ravi Roy
NETWORKS Guido Caldarelli and Michele Catanzaro
THE NEW TESTAMENT Luke Timothy Johnson
THE NEW TESTAMENT AS LITERATURE Kyle Keefer
NEWTON Robert Iliffe
NIELS BOHR J. L. Heilbron
NIETZSCHE Michael Tanner
NINETEENTH-CENTURY BRITAIN Christopher Harvie and H. C. G. Matthew
THE NORMAN CONQUEST George Garnett
NORTH AMERICAN INDIANS Theda Perdue and Michael D. Green
NORTHERN IRELAND Marc Mulholland
NOTHING Frank Close
NUCLEAR PHYSICS Frank Close
NUCLEAR POWER Maxwell Irvine
NUCLEAR WEAPONS Joseph M. Siracusa
NUMBER THEORY Robin Wilson
NUMBERS Peter M. Higgins
NUTRITION David A. Bender
OBJECTIVITY Stephen Gaukroger
OCEANS Dorrik Stow
THE OLD TESTAMENT Michael D. Coogan
THE ORCHESTRA D. Kern Holoman
ORGANIC CHEMISTRY Graham Patrick
ORGANIZATIONS Mary Jo Hatch
ORGANIZED CRIME Georgios A. Antonopoulos and Georgios Papanicolaou
ORTHODOX CHRISTIANITY A. Edward Siecienski
OVID Llewelyn Morgan
PAGANISM Owen Davies
PAIN Rob Boddice
THE PALESTINIAN-ISRAELI CONFLICT Martin Bunton
PANDEMICS Christian W. McMillen
PARTICLE PHYSICS Frank Close
PAUL E. P. Sanders
PEACE Oliver P. Richmond
PENTECOSTALISM William K. Kay
PERCEPTION Brian Rogers
THE PERIODIC TABLE Eric R. Scerri
PHILOSOPHICAL METHOD Timothy Williamson
PHILOSOPHY Edward Craig
PHILOSOPHY IN THE ISLAMIC WORLD Peter Adamson
PHILOSOPHY OF BIOLOGY Samir Okasha
PHILOSOPHY OF LAW Raymond Wacks
PHILOSOPHY OF SCIENCE Samir Okasha
PHILOSOPHY OF RELIGION Tim Bayne
PHOTOGRAPHY Steve Edwards
PHYSICAL CHEMISTRY Peter Atkins
PHYSICS Sidney Perkowitz
PILGRIMAGE Ian Reader
PLAGUE Paul Slack
PLANETS David A. Rothery
PLANTS Timothy Walker
PLATE TECTONICS Peter Molnar
PLATO Julia Annas
POETRY Bernard O'Donoghue
POLITICAL PHILOSOPHY David Miller
POLITICS Kenneth Minogue
POPULISM Cas Mudde and Cristóbal Rovira Kaltwasser
POSTCOLONIALISM Robert Young
POSTMODERNISM Christopher Butler
POSTSTRUCTURALISM Catherine Belsey
POVERTY Philip N. Jefferson
PREHISTORY Chris Gosden
PRESOCRATIC PHILOSOPHY Catherine Osborne
PRIVACY Raymond Wacks
PROBABILITY John Haigh
PROGRESSIVISM Walter Nugent
PROHIBITION W. J. Rorabaugh
PROJECTS Andrew Davies
PROTESTANTISM Mark A. Noll
PSYCHIATRY Tom Burns
PSYCHOANALYSIS Daniel Pick
PSYCHOLOGY Gillian Butler and Freda McManus

PSYCHOLOGY OF MUSIC Elizabeth Hellmuth Margulis
PSYCHOPATHY Essi Viding
PSYCHOTHERAPY Tom Burns and Eva Burns-Lundgren
PUBLIC ADMINISTRATION Stella Z. Theodoulou and Ravi K. Roy
PUBLIC HEALTH Virginia Berridge
PURITANISM Francis J. Bremer
THE QUAKERS Pink Dandelion
QUANTUM THEORY John Polkinghorne
RACISM Ali Rattansi
RADIOACTIVITY Claudio Tuniz
RASTAFARI Ennis B. Edmonds
READING Belinda Jack
THE REAGAN REVOLUTION Gil Troy
REALITY Jan Westerhoff
RECONSTRUCTION Allen C. Guelzo
THE REFORMATION Peter Marshall
RELATIVITY Russell Stannard
RELIGION Thomas A. Tweed
RELIGION IN AMERICA Timothy Beal
THE RENAISSANCE Jerry Brotton
RENAISSANCE ART Geraldine A. Johnson
RENEWABLE ENERGY Nick Jelley
REPTILES T. S. Kemp
REVOLUTIONS Jack A. Goldstone
RHETORIC Richard Toye
RISK Baruch Fischhoff and John Kadvany
RITUAL Barry Stephenson
RIVERS Nick Middleton
ROBOTICS Alan Winfield
ROCKS Jan Zalasiewicz
ROMAN BRITAIN Peter Salway
THE ROMAN EMPIRE Christopher Kelly
THE ROMAN REPUBLIC David M. Gwynn
ROMANTICISM Michael Ferber
ROUSSEAU Robert Wokler
RUSSELL A. C. Grayling
THE RUSSIAN ECONOMY Richard Connolly
RUSSIAN HISTORY Geoffrey Hosking
RUSSIAN LITERATURE Catriona Kelly
THE RUSSIAN REVOLUTION S. A. Smith
SAINTS Simon Yarrow
SAVANNAS Peter A. Furley
SCEPTICISM Duncan Pritchard
SCHIZOPHRENIA Chris Frith and Eve Johnstone
SCHOPENHAUER Christopher Janaway
SCIENCE AND RELIGION Thomas Dixon
SCIENCE FICTION David Seed
THE SCIENTIFIC REVOLUTION Lawrence M. Principe
SCOTLAND Rab Houston
SECULARISM Andrew Copson
SEXUAL SELECTION Marlene Zuk and Leigh W. Simmons
SEXUALITY Véronique Mottier
SHAKESPEARE'S COMEDIES Bart van Es
SHAKESPEARE'S SONNETS AND POEMS Jonathan F. S. Post
SHAKESPEARE'S TRAGEDIES Stanley Wells
SIKHISM Eleanor Nesbitt
THE SILK ROAD James A. Millward
SLANG Jonathon Green
SLEEP Steven W. Lockley and Russell G. Foster
SMELL Matthew Cobb
SOCIAL AND CULTURAL ANTHROPOLOGY John Monaghan and Peter Just
SOCIAL PSYCHOLOGY Richard J. Crisp
SOCIAL WORK Sally Holland and Jonathan Scourfield
SOCIALISM Michael Newman
SOCIOLINGUISTICS John Edwards
SOCIOLOGY Steve Bruce
SOCRATES C. C. W. Taylor
SOFT MATTER Tom McLeish
SOUND Mike Goldsmith
SOUTHEAST ASIA James R. Rush
THE SOVIET UNION Stephen Lovell
THE SPANISH CIVIL WAR Helen Graham
SPANISH LITERATURE Jo Labanyi
SPINOZA Roger Scruton
SPIRITUALITY Philip Sheldrake
SPORT Mike Cronin
STARS Andrew King
STATISTICS David J. Hand
STEM CELLS Jonathan Slack
STOICISM Brad Inwood
STRUCTURAL ENGINEERING David Blockley

STUART BRITAIN John Morrill
THE SUN Philip Judge
SUPERCONDUCTIVITY Stephen Blundell
SUPERSTITION Stuart Vyse
SYMMETRY Ian Stewart
SYNAESTHESIA Julia Simner
SYNTHETIC BIOLOGY Jamie A. Davies
SYSTEMS BIOLOGY Eberhard O. Voit
TAXATION Stephen Smith
TEETH Peter S. Ungar
TELESCOPES Geoff Cottrell
TERRORISM Charles Townshend
THEATRE Marvin Carlson
THEOLOGY David F. Ford
THINKING AND REASONING Jonathan St B. T. Evans
THOMAS AQUINAS Fergus Kerr
THOUGHT Tim Bayne
TIBETAN BUDDHISM Matthew T. Kapstein
TIDES David George Bowers and Emyr Martyn Roberts
TOCQUEVILLE Harvey C. Mansfield
TOPOLOGY Richard Earl
TRAGEDY Adrian Poole
TRANSLATION Matthew Reynolds
THE TREATY OF VERSAILLES Michael S. Neiberg
TRIGONOMETRY Glen Van Brummelen
THE TROJAN WAR Eric H. Cline
TRUST Katherine Hawley
THE TUDORS John Guy
TWENTIETH-CENTURY BRITAIN Kenneth O. Morgan
TYPOGRAPHY Paul Luna
THE UNITED NATIONS Jussi M. Hanhimäki
UNIVERSITIES AND COLLEGES David Palfreyman and Paul Temple
THE U.S. CIVIL WAR Louis P. Masur
THE U.S. CONGRESS Donald A. Ritchie
THE U.S. CONSTITUTION David J. Bodenhamer
THE U.S. SUPREME COURT Linda Greenhouse
UTILITARIANISM Katarzyna de Lazari-Radek and Peter Singer
UTOPIANISM Lyman Tower Sargent
VETERINARY SCIENCE James Yeates
THE VIKINGS Julian D. Richards
VIRUSES Dorothy H. Crawford
VOLCANOES Michael J. Branney and Jan Zalasiewicz
VOLTAIRE Nicholas Cronk
WAR AND TECHNOLOGY Alex Roland
WATER John Finney
WAVES Mike Goldsmith
WEATHER Storm Dunlop
THE WELFARE STATE David Garland
WILLIAM SHAKESPEARE Stanley Wells
WITCHCRAFT Malcolm Gaskill
WITTGENSTEIN A. C. Grayling
WORK Stephen Fineman
WORLD MUSIC Philip Bohlman
THE WORLD TRADE ORGANIZATION Amrita Narlikar
WORLD WAR II Gerhard L. Weinberg
WRITING AND SCRIPT Andrew Robinson
ZIONISM Michael Stanislawski

Available soon:

AMPHIBIANS T. S. Kemp
MODERN BRAZIL Anthony W. Pereira
WAR AND RELIGION Jolyon Mitchell and Joshua Rey
MARKETING Kenneth Le Meunier-FitzHugh
THE VIRTUES Craig A. Boyd and Kevin Timpe

For more information visit our website

www.oup.com/vsi/

Michael J. Branney and Jan Zalasiewicz

# VOLCANOES

## A Very Short Introduction

OXFORD
UNIVERSITY PRESS

Great Clarendon Street, Oxford, OX2 6DP,
United Kingdom

Oxford University Press is a department of the University of Oxford.
It furthers the University's objective of excellence in research, scholarship,
and education by publishing worldwide. Oxford is a registered trade mark of
Oxford University Press in the UK and in certain other countries



First edition published in 2021

Impression: 1



Published in the United States of America by Oxford University Press
198 Madison Avenue, New York, NY 10016, United States of America

British Library Cataloguing in Publication Data
Data available

Library of Congress Control Number: 2020944347

ISBN 978–0–19–958220–4

Printed in Great Britain by
Ashford Colour Press Ltd, Gosport, Hampshire

*For Malcolm Howells and Tony Reedman:*
*inspirational investigators of the ancient*
*volcanoes of Snowdonia*

# Contents

# Prologue

When Mount Pinatubo in the Philippines erupted in 1991 the landscape was transformed. Before, indigenous Aeta tribesmen had hunted and gathered there amongst mature tropical forest, wildlife, and verdant slopes. Now their homes and the forest were gone, replaced by a dusty plateau of gleaming white pumice and ash. Somewhere far below lay the original landscape: torched and deeply buried, the communities that lived on it ripped apart, their way of life obliterated. The most powerful volcanic eruption of the century to affect an inhabited landscape had just taken place.

Some time later a team of Filipino, American, and English volcanologists arrived by helicopter at that desolate terrain, driven by curiosity, the fascination of puzzling out what happened, and the sense that detailed understanding of this cataclysm could be useful when the next eruption threatens. There was practicality in that, too: a volcano is not the best place to be when a major eruption is in full swing, and little can be seen anyway in the opaque, ash-choked atmosphere. So it is preferable, and considerably safer, to visit the site after the eruption has ceased, and then attempt to reconstruct the cataclysmic events forensically, from evidence within the various layers of ash and rubble deposited during the eruption. Leave the site a couple of years, and streams will have cut gullies through the soft deposits, revealing the underlying layers, so that the hidden chapters of events, such as the opening phases of the eruption, and any precursor activity, might also be deciphered.

Geologists are drawn to such fieldwork not merely for love of the outdoors, but because of the enjoyment and satisfaction one gets from framing and trying to solve a myriad related hypotheses on the hoof. What could have formed that deposit over there? Why is it absent here? What smashed that? What happened to that soil? Careful documentation of the complex architectures of volcanic strata, such as the placing of one piece of debris relative to another, can lead to unexpected deductions about what really occurred, often aided by a variety of laboratory tests on collected samples. Field expertise is a craft acquired during long experience working on and around several volcanoes, so that a volcanologist may quickly spot which features at a site are unusual or significant. By assembling evidence to answer the questions posed (often by systematically disproving competing ideas of what happened), one can reconstruct the rapid-fire series of events that no living person could ever have witnessed. And from many such studies, more general truths can emerge about how volcanoes behave.



Volcanology is a young science, and many new discoveries and insights remain to be made. Indeed, in recent decades each major eruption (Mount St Helens, Kilauea, Pinatubo, Montserrat) has yielded a step-change in volcanological understanding. The idea of a volcano as a perfect cone-like peak, intermittently flaming into life amid floods of magma and showers of incandescent ash, no longer holds true as the norm. Volcanologists today have a working mental library of much more diverse forms and processes.

Many volcanoes, for instance, are largely gigantic sheets of rubble, following rapid collapse of spectacular but unstable, and therefore relatively short-lived, cone-shaped peaks. Others—and these tend to be among the most powerful and dangerous ones—are vast and deceptively peaceful depressions in the ground, so subdued as sometimes to be omitted from volcano maps. And some of the most innocuous-looking ash layers are now known to represent some of the most cataclysmic eruptions. Some iconic images—for instance of fiery floods of lava racing to engulf entire landscapes—have been replaced by an appreciation of far more intricate, and slower, lava flow behaviour. And as astronomers begin to explore the geology of other planets and moons, they are finding yet more surprises—of volcanoes larger and more ancient than any on Earth, and yet others built of ice.

In this book, we provide a flavour of how volcanoes and volcanism, on Earth and far beyond, are understood today. Our account is neither technical nor comprehensive. For such deeper study, there are excellent textbooks that we recommend. Instead we aim to provide a brief but vivid introduction to the subject, to some of the important processes involved, and to the approaches adopted by volcanologists as they attempt to learn more of one of the Earth's most distinctive and dramatic phenomena. We still have lots to learn—be assured that future eruptions will continue to both surprise and fascinate.

We would like to thank Jenny Nugee, Latha Menon, Chandrakala Chandrasekaran and their colleagues at OUP, for their constant support for and patience with this project, and Alan Rose for reviewing the manuscript. Our deep thanks too, to Brian Bell, Peter Kokelaar, Magnús Guðmundsson, and John Smellie for providing figures, and to our many colleagues and mentors over the years, from whom we have absorbed something of the science and craft of field volcanology.

# List of illustrations

# Chapter 1
# The making of magma

## Heating a mobile Earth

Volcanism is driven by prodigious amounts of heat from deep inside the Earth that partly melts rock to form magma, and drives powerful eruptions that expel ash (small fragments of solidified magma) and lava on to the surface, creating what we call a volcano. The source of that heat has been puzzled over since the times of antiquity, but we now know its main source is a phenomenon that was long unsuspected. When radioactivity—the random splitting of atomic nuclei to generate new elements, with the release of radiation—was discovered, by Henri Becquerel in 1897, it was quickly recognized as the source of heat and energy that could drive a dynamic Earth. Even the low levels of radioactivity deep inside our planet (mantle levels are lower than in the crust) are sufficient for heat to build to the point at which rocks locally begin to melt.

This heat engine also causes the crust to move so that mountains form and powerful earthquakes are generated. The sheer scale of that motion was not fully appreciated until it was seen that the continents themselves were drifting over the face of the Earth: eventually—after many millions of years—moving thousands of miles. This idea was promoted spiritedly by the perceptive explorer and scientist Alfred Wegener in the early 20th century—and was generally dismissed by geophysicists, for want of a mechanism to drive these enormous crustal masses through the oceans. But Wegener's observation that the rocks of western Africa closely matched those of eastern America, and the inference from this that the continents had once lain together, side-by-side, remained sound. Then, in the 1960s, it was discovered that the rock beneath the deep oceans was geologically young—much younger than the continents—and was being constantly created at mid-ocean ridges and then destroyed at ocean trenches. In this revolutionary new paradigm of plate tectonics, the continents were simply carried along on the drifting plates. Being too buoyant to sink back into the Earth's mantle, they just travelled on, growing ever more ancient.

Plate tectonics is the fundamental controller and context for pretty well all Earth surface processes—not least volcanism (Figure 1). The mid-ocean ridges themselves might be thought of as enormously long, planetary-scale volcanoes. Eruptions take place along the continuously opening fissure at the ridge crest, adding more material to the growing edges of the ocean plates. But volcanoes also arise where the ocean crust sinks back into the depths of the mantle, typically in a volcanic arc, far above the inclined, descending ocean plate, within a few hundred kilometres of the ocean trench. Examples of arc volcanoes include the mighty Citlaltépetl, the highest peak in Mexico, Mount Fuji, the highest peak in Japan, the majestic Cotopaxi and Chimborazo in Ecuador, and Mount Taranaki, which towers over North Island, New Zealand.

Volcanoes also arise far away from mid-ocean ridges and arcs, above 'hot spots'. These hot spots are thought to mark the location of slowly ascending subterranean plumes of hot, solid mantle material, perhaps 1000 kilometres high and hundreds of kilometres across. Above these plumes, the Earth's crust and uppermost mantle move along like a conveyor belt above a

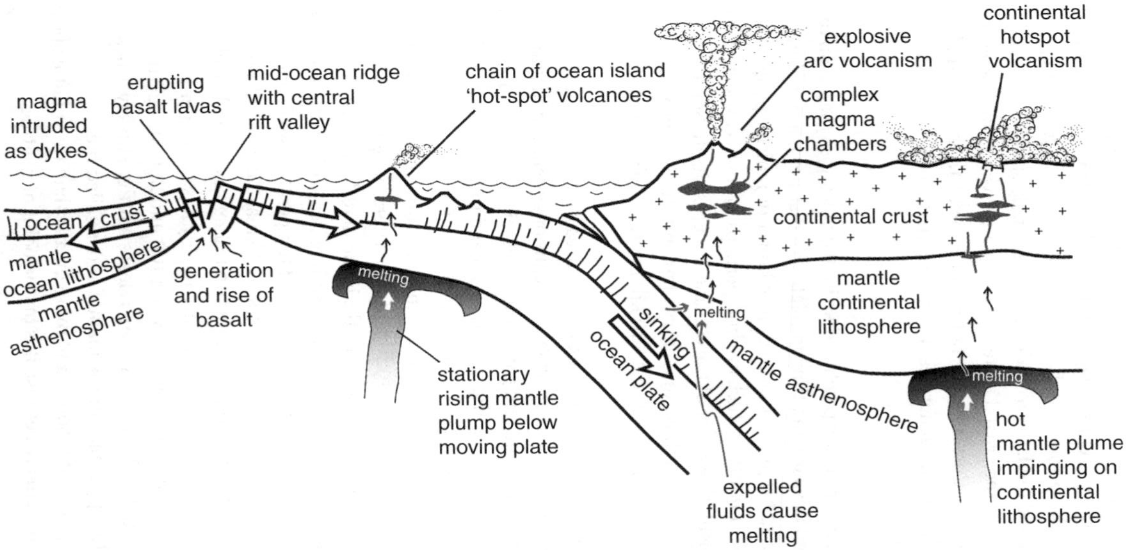

1. **Plate tectonic setting of the main kinds of volcanism on Earth (vertically exaggerated, not to scale).**

stationary blowtorch, and melting begins where the blowtorch is brought to bear. The classic example is the chain of Hawaiian islands, where volcanoes have successively grown, and then faltered and died, to be eroded and eventually sink beneath the waves when the crustal plate moved on and away from the active hot-spot site. Currently, the hot-spot lies beneath the island of Hawaii itself, and generates the magma that feeds its copious lava flows.

Then there are more patchily distributed volcanoes that seem too dispersed in space and time to relate to a distinct mantle plume. These might represent more diffuse hotter or moister patches of the upper mantle—which are hot, perhaps in some cases, because of the insulating effect of the continental crust itself. Some magmas, such as those that carry diamonds to the surface, come from much deeper, perhaps 500 kilometres down and more.



Each of these volcanic settings—mid-ocean ridges, arcs, and hot-spots—has its own particular means of generating magma; that is, of producing and gathering together sufficient amounts of molten rock material that may then ascend to the Earth's surface. This process needs special conditions. The Earth is not made of a thin solid crust precariously floating on a subterranean sphere of boiling magma, as some 19th-century writers suggested. The crust is solid, to be sure—but so, mostly, is the mantle. For, while temperatures steadily rise on descending deeper into the Earth, so do pressures. And although increased temperatures will encourage a rock to melt, increased pressures act to make it remain solid, even at high temperatures. These two influences vie with each other deep into the Earth. Pressure mostly has the upper hand—except, seemingly, in a layer between 100 and 200 kilometres down in the Earth, where its grip relaxes a little. This layer lies within the upper part of the mantle and is called the asthenosphere. It is softer and more ductile than the mantle above and below it, likely because it is on the point of melting.

(The more rigid mantle above it, together with the overlying crust, form the Earth's tectonic plates, or 'lithosphere'.)

At mid-ocean ridges, the pulling apart of the ocean plates relaxes the pressure on the mantle, and as a result the asthenosphere comes a little closer to the surface and enhanced melting takes place. The magma produced by this melting, though, does not have the same composition as the mantle. The mantle rocks are mainly made of dense minerals, richer in iron and magnesium and lower in silica and aluminium than most rocks we see at the surface. When they *just* begin to melt, the few per cent of the rock that first melts is richer in silica than the solid residue left behind, which is poorer in silica and more resistant to melting.

Thus is formed *basalt* magma (about 53 per cent silica) from a source material that is generally much denser (less than 45 per cent silica). This partial melting process is astonishingly uniform, across all of the world's mid-ocean ridges, whether they are long or short, and whether the ocean plates are drifting apart slowly or more rapidly. Consequently almost all of the ocean crust is made of basalt. Not only that, it is a type of basalt that is subtly and chemically distinct from, say, basalt that erupts from a hot-spot, as at Hawaii. The Earth machine, in some respects, is *very* well regulated.

The volcanoes that erupt above subduction zones are quite different, and their origin is equally distinct, involving not so much heat or pressure, as water. It is a somewhat counter-intuitive process, for the ocean crust that sinks deep into the mantle is dense, old, and cold, and it cools the mantle it comes into contact with. But it also carries with it water from the oceans, and this water is released into the mantle above the descending ocean plate. This water lowers the melting point of the mantle so melting results. As with magma generated at mid-ocean ridges, the small fraction of melt produced has a different composition to the solid material from which it is derived.

## Gathering magma

The accumulation of magma takes place deep underground, hidden from our eyes. The process can only be studied by inferences based upon volcano behaviour, from seismic information, chemical tracers in volcanic materials, and other clues. Many mysteries therefore remain—and one of these is how the magma slowly gathers in the first place, from tiny, scattered patches of melt (that comprise only a few per cent of the source mantle rock) into large, better-defined underground magma reservoirs and conduits.

The problem is this: 100 kilometres or so below the Earth's surface, some hundreds to thousands of cubic kilometres of mantle rock are just at the point of melting. Finely dispersed within this enormous volume is a delicate three-dimensional skein of magma, in diaphanous sheets and thin strands and droplets. How does this then find its way and move across tens to hundreds of kilometres towards the volcano?

There is a driving force—gravity. The melted material (relatively rich in light elements such as silicon) is less dense than the remaining solid matter (rich in heavier elements such as iron and magnesium). So, the melt will always try to rise relative to the solid. Set against that is the friction of the tiny melt droplets against the solid crystalline maze that they trickle through, and the tortuousness and sheer distance of the path they must travel. So—one answer might be... *slowly*. At many volcanoes the repose time between eruptions can be enormous: at Tenerife major, island-devastating eruptions were each separated by several thousands to hundreds of thousands of years. Over time (and most likely even now), magma is, infinitesimally slowly, gathering towards the major storage reservoirs from which—at some critical point—rapid ascent can begin.

Elsewhere, though, the magma seems to percolate rather more rapidly through the mantle. The lavas that erupt from mid-ocean ridges, for instance, contain short-lived radioactive isotopes (of elements such as protoactinium and radium). Their survival to the surface suggests that magma can migrate upwards through the mantle at speeds approaching 100 metres per year. Is this possible?

One can experiment, and make a mantle in microcosm in the laboratory. Crystals of olivine (as solid mantle material) can be mixed with small amounts of basalt (the melt), pressurized in a piston, and heated *just* to melting—some 1300°C. The resultant 'mantle' is then spun rapidly in a centrifuge (to intensify the effect of gravity, as scientists need results in hours rather than millennia!). The 'magma' in such an experiment travels through the crystal mush at speeds equivalent to those deduced for the mid-ocean ridge basalts. Artifice here has, it seems, successfully reproduced nature.

Magma travel through subterranean rock may be speeded up by other processes. In the laboratory, take a crystal-rich mush with a small amount of melt dispersed throughout it and then shear that material—as might naturally happen, say, under the deforming roots of a mountain belt. The melted material separates out into systems of parallel planar sheets, less than a millimetre thick and inclined at an angle of some 20 degrees to the plane of shear. Once the melt is concentrated within such sheets, it can travel faster than as a dilute dispersion.

From such enormous, diffuse networks deep in the Earth, the magma rises and gathers together in the crust as a complex system of interconnected fractures, reservoirs, and conduits.

These may act as holding centres—perhaps for tens of thousands of years.

## Probing magma reservoirs

Magma reservoirs beneath a volcano are difficult to study; they lie deep below ground, and are not directly accessible. But something of their size and depth can sometimes be sensed, using indirect clues. The chemistry of minerals carried up to the surface by a volcanic eruption may betray the pressures (i.e. depth) and temperature at which the magma in the reservoir crystallized. And, geophysicists can 'echo-sound' the rock structure beneath a volcano, by seismic imaging. This works by setting off small controlled explosions and then measuring the resultant shock waves after they have passed beneath the volcano and variously bounced back to the surface. Magma reservoirs can be detected by the way they slow these man-made seismic waves.



This kind of study has been made of Mount St Helens, the most frequently active volcano in north America. Mount St Helens has erupted seven times in the last 4000 years, the last time in 1980, with fatal consequences for fifty-seven people. Seismic imaging has shown a 'low-velocity' zone about 10 kilometres in diameter beneath the volcano, offset a little to the south and west of the summit and extending from a depth of about 4 to 14 kilometres. Similar depths have been deduced from the chemistry of the crystals erupted out of Mount St Helens over the past 4000 years. This suggests that the low-velocity zone represents a long-lived magma reservoir that has fed successive eruptions over the millennia. The magmas must get through a barrier, though, shown by a *high* velocity zone just a couple of kilometres below the volcano, and thought to represent an ancient body of magma that cooled underground into a rigid mass. This barrier may have prevented activity in 2004 and 2008 developing into

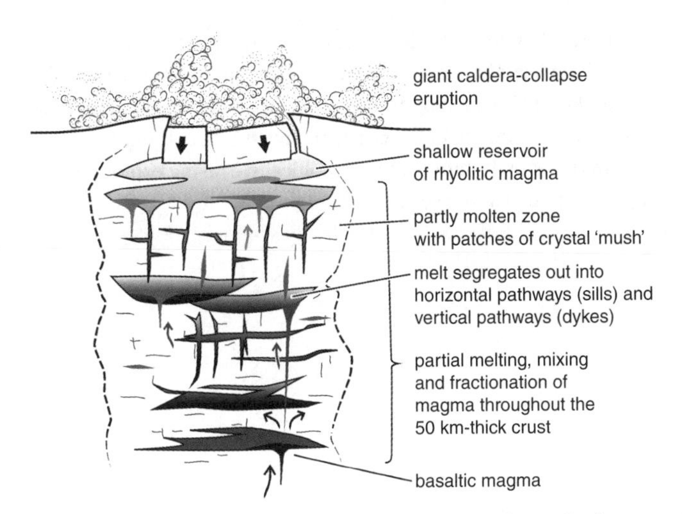


**2. Magma collection and crustal reservoir formation beneath a large continental volcano.**



eruptions, a factor to be considered when forecasting future eruptions.

Mount St Helens is but a single and modestly sized volcano. Larger volcanoes have magma chambers in proportion (Figure 2). Long Valley caldera volcano in California is the source of the three-quarters-of-a-million-years-old 'Bishop Tuff', an outpouring of ash ('tuff' is its hardened version) that reached the prodigious volume of 500 cubic kilometres, and affected much of north America. Here, seismic imaging has vividly picked out a melt-rich zone beneath the caldera that extends through much of the crust. This zone seems to have two parts: a more silica-rich melt above and silica-poor basaltic melt below, and has received magma since the last major eruption, 300,000 years ago. There seems to be enough magma down there for another huge, Bishop Tuff-like eruption—though, fortunately, that is not an indication that another such cataclysm is imminent. A current preoccupation of volcanologists is what conditions allow large underground magma reservoirs to stagnate in the crust, and what causes them at certain times to destabilize catastrophically, causing explosive super-eruptions at the surface.

# Chapter 2
# How do volcanoes explode?

## Anatomy of an explosive eruption

A few kilometres beneath an active volcano lies a subterranean magma reservoir, in which numerous interconnected fractures and fissures are filled with partly liquid magma. This magma initially has the temperature of a bonfire—around 1100°C—but gradually loses heat to the surrounding rock. As it cools, crystals begin to grow, changing the chemical composition of the residual liquid. As the crystals grow more abundant, the magma eventually turns into a hot sticky mush of crystals, from which the hot liquid (by this time of a quite different composition to the original magma) may ooze out, rather like water oozes out of disturbed quicksand.

Meanwhile, the rock walls of the subterranean fractures heat up, in some cases sufficiently to start melting. Fragments of rock detach from the fracture walls and float out into the magma where they, too, begin to melt and become assimilated, further changing the overall chemical composition of the magma.

Hotter, new magma with a different composition may rise from depth. On arrival in the reservoir it mingles and stirs with the earlier, partly cooled and now more viscous, mushy magma, forming streaks and bands that resemble the intricate marbling of raspberry ripple ice-cream. If the two magmas are equally runny they mingle easily and, with a little more stirring, they blend together. But if one magma is less runny than the other, they do not mix together so readily.

As a result of these three processes—crystallization, assimilation, and mixing—the magma beneath a volcano is constantly changing, and this may continue over a long time. The location, including depth, of the magma also may change with time.

Gases such as carbon dioxide, sulphur dioxide, and steam are naturally present within hot magma. Perturb the magma in some way—for example by introducing a second, hotter batch of magma, by repacking all those loosely arranged crystals so that the interstitial liquid wells out, or by dropping the pressure—and the dissolved gases may start to come out of solution, forming bubbles. For evidence of this, examine a piece of pumice from a volcano (try your local pharmacy): it appears pale, and lightweight. Through a microscope, it can be seen to be made of shiny volcanic glass. Volcanic glasses (obsidians) are commonly

**3. Agathla Peak in Arizona is the exhumed conduit of a 25–30 million year-old extinct volcano. Hardened magma and breccia tower above softer-weathering sandstones of the surrounding plain.**

dark because they contain nano-scale inclusions of dark minerals such as magnetite. These occur in pumice too, but pumice also contains millions of microscopic bubbles. It is a solidified microfoam, like polystyrene or meringue: light enough to float on water because it is mostly trapped air, and pale because of the play of light upon the myriad tiny bubbles (too small to see individually with the naked eye) and their translucent, gossamer-thin bubble-walls. In a similar way, the microfoam head that floats on your pint of beer is both paler and less dense than the darker liquid beneath.

## Pumice-forming rhyolitic eruptions

Most pumice is silica-rich. Silica-rich (rhyolitic) magmas tend to be viscous, with a consistency closer to warm toffee than runny water. This is because their internal structure is highly polymerized, with large, intertwined silicate molecules that don't readily move past each other. Dissolved gases, such as carbon dioxide and water, can move through magma by diffusion. But diffusion through a sticky, highly polymerized magma is very slow. If it takes too long for dissolved gas to diffuse to the nearest bubble, the gas may instead simply cause the nucleation of a new, even nearer bubble. So in silicic magma, millions of very tiny bubbles nucleate, creating a microfoam.

As the magma froths up, it becomes more buoyant and so starts to ascend, usually up a crack in the rock, which funnels the magma and is therefore known as a 'conduit'. Exactly what causes the crack or conduit initially to propagate through solid rock at a particular site remains rather poorly understood. As the rising magma reaches shallower levels beneath the volcano, the pressure continues to drop, so it froths even more, and consequently rises ever faster. This in turn further reduces the pressure, causing more frothing and ever more acceleration up the conduit, and by the time it exits the volcano the foaming magma is moving at supersonic speed. But it is far too viscous to flow this rapidly, so at some level within the conduit, it rips apart like the snapping of rapidly pulled-apart elastic chewing gum. What finally exits the volcano is a jet of incandescent fragments of magma froth-supported in explosively expanding hot gas. Most people are familiar with explosions that go bang, such as a cannon shot or bomb burst. Such explosions are short in duration because the fuel is rapidly spent within a fraction of a second. In a volcano, however, the volume of fuel (the magma) is far greater—several cubic *kilometres* of magma. Even as the first material is explosively ejected from the volcano, more frothing, fragmenting, and expanding magma follows it close behind. So the explosion is *sustained*, and it roars like a super-sized rocket turbine. The roar may last for several hours before the volcanic explosion finally ceases.

Cracked open, a piece of pumice appears fibrous, with an internal grain as in wood. Through a microscope one can see that each bubble is highly elongate, rather than spherical, and aligned parallel to adjacent bubbles. The bubbles evidently stretched while the foam was still hot and malleable (Figure 4). Yet the individual fragments of pumice are not themselves so elongate,



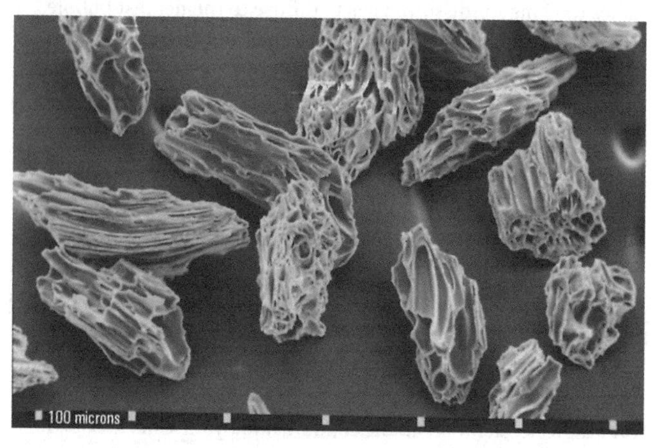


**4. A scanning electron microscope image of pumice fragments with stretched micro-bubbles. The Rockland Ash Bed, Cascades, USA.**

and therefore the ductile stretching must have happened before the breakage. It appears that the frothing magma was stretching while it was rising up the volcano's conduit but below the level where the flow was sufficient to fragment it. By the time the foam fragmented, it had become too rigid for the bubbles to return to a spherical shape, and so the distinctive elongate pumice texture is frozen evidence of its earlier underground journey.

Pieces of pumice form white layers that extend over 30 kilometres from an explosive volcano, blanketing the entire landscape, covering hills and valleys alike. The layers can be seen in road cuts around Naples in Italy, in central Mexico, Japan, Santorini, Tenerife, and elsewhere. Each layer is a loosely packed mass of angular pumice fragments: the fragments touch one another at point contacts but with abundant airspace in between (Figure 5). Scattered amongst the pumice are smaller fragments of different types of rock: lava,

**5. A Plinian pumice fall deposit, with angular fragments of pumice (pale) and older rock (grey) from the conduit walls. The fragments are angular with open spaces between. Fine ash blew farther downwind.**

granite, even limestone and sandstone. These are little bits of the volcano and bedrock that were blown up. They are smaller than the enclosing pumice fragments because they are denser, and settle faster, closer to source. Overall the fragments are extraordinarily well sorted: the largest fragments are found close to the volcano, and the size of the fragments overall decreases systematically away from the volcano. A hundred kilometres from the volcano the layer is much thinner and of dust-sized particles.

The pieces of pumice drop from a staggering 20–50 km altitude, well up into the stratosphere. How do they get that high? Imagine firing a cannonball straight upwards. It would take an enormous cannon to fire to just 5 kilometres height, and it is virtually impossible to fire a ball to an altitude of tens of kilometres ballistically—the volcanic jet from a powerful exploding volcano only reaches a kilometre or so above the volcano's summit. The key to the extraordinary vertical transport is that the volcanic jet is very hot, and sucks surrounding cold atmospheric air into it. The cold air is entrained into the erupting jet and is heated rapidly by the hot pumice and ash particles, so that it expands. The thermal expansion is dramatic, and if sufficient atmospheric air is entrained into the edges of the erupting mixture, the entire hot, turbulent mixture of pumice, ash, rock fragments, air, and volcanic gases becomes less dense than the surrounding atmosphere. It then lofts buoyantly, like the smoke rising above a conflagration, only much faster—fast enough, indeed, to carry cannonball-sized rocks up to the stratosphere. The volcanic plume continues to rise until it reaches an altitude where the surrounding atmosphere is so rarefied that its density matches that of the plume. This is known as the level of neutral buoyancy. The altitude that the eruption plume attains reflects the thermal energy supplied by the volcano within a given period of time: the larger the flux of hot ejecta from the volcano, the higher the resultant eruption column.

But it doesn't stop there: more material rises close behind, shoving the top of the volcanic column aside so that the upper part starts to spread outwards like a giant opening umbrella (Figure 6). This 'umbrella cloud' rolls out across the sky in all directions, blocking out sunlight to the district below like a monstrous window blind. And it is from here that the pumice begins to rain down onto the landscape, far below.

This type of eruption is known as a 'Plinian' eruption, after Pliny the Younger, who famously described such an eruption at Vesuvius in 79 AD. Plinian eruptions can be immense: during the 1991 Plinian eruption of Mount Pinatubo in the Philippines a wide region of Luzon was blanketed under a layer of pumice. City-sized military bases were evacuated just in time, and over half a million inhabitants left their homes. In just three and a half hours, between 5 and 8 cubic kilometres of new magma were ejected, and among its wider effects was a measurable change in world climate for a couple of years.

You won't die if a few centimetre-sized pieces of pumice drop on your head, although the accompanying dense rock fragments may

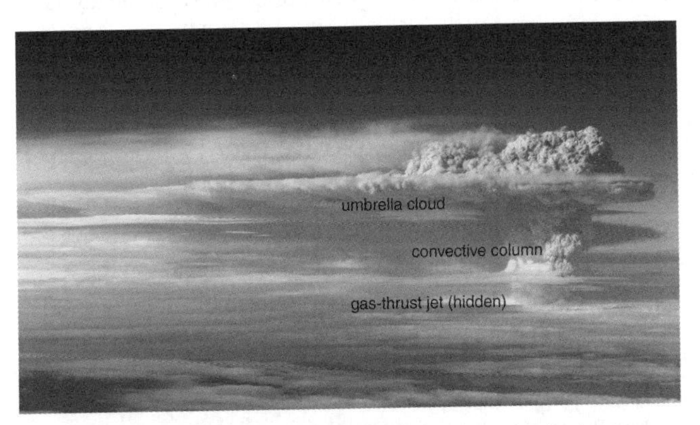


**6. A Plinian eruption column from Puyehue volcano, Chile, in 2011.**

be more painful. It's a bit like being caught in a heavy hailstorm when folk rush for shelter, and cars acquire pockmarked, dinted roofs. But hail rarely accumulates to more than a centimetre or two. Snowfall can be thicker: layers of just a few tens of centimetres bring British cities to a halt until they melt. However, pumice and ash don't melt: they remain, clogging roof gutters, drains, watercourses, and sewerage for days, weeks, and years. Volcanic ash blows into drifts, buries gardens and crops. Try to remove it from your car and it scratches both paintwork and windscreen—unlike wood ash it is highly abrasive. It may not cause too much damage accumulating on a house or warehouse roof—until it rains, when it soaks up the rainwater, the considerable extra weight often causing flat roofs to collapse, destroying homes and commercial buildings. In rural areas travel is made difficult, and food is in short supply. Crops and livestock die.



The Pinatubo pumice layer that caused so much destruction was only about 20 centimetres thick at the nearby Clarke airbase. However, pumice layers elsewhere reach much greater thicknesses, such as the metre-thick layer that buried the ancient Minoan port of Akrotiri on Santorini, or the prehistoric pumice fallout layers of over 8 metres thickness on Tenerife, even 10 kilometres distant from the volcano.

## Deciphering previous explosive events

One of the first things a volcanologist may do on arrival at a volcano is to figure out how the volcano has behaved in the past: what sort of eruptions, how large, and how frequent. All the required information is there on the ground in the form of a series of layers of ash and pumice, but it can take quite a bit of know-how and several days to explore and puzzle out what it all means. Although pumice layers do not melt, they can become wind-blown and washed away after rainfall by streams, so the record of previous eruptions is often less complete than one would wish.

Furthermore, whilst some of the eruptions of a volcano may be recorded at one site, other eruptions are recorded at different sites, so to reconstruct the history of eruptions, all the evidence needs very careful piecing together like a giant jigsaw puzzle. Thinner ash layers are particularly prone to removal, and are sometimes best preserved where there are no streams, such as within tranquil lakes, in slowly forming peat bogs, or within layers of snow and ice. The pumice layer from the 1875 Askja eruption in central Iceland can still be found on top of a pre-1875 layer of snow that, remarkably, has not yet completely melted.

Most volcanoes have erupted numerous times, leaving many layers of pumice (Figure 7). Each layer must then be somehow distinguished from the others: one layer may have a particularly distinctive colour or composition, or display a distinctive pattern of grain size layering, rather like a fingerprint. It is therefore fortunate that any patterns of internal layering, such as gradual or abrupt changes in particle size with height, remain constant

**7. Parallel fallout layers of white Plinian pumice and dark basaltic scoria. Ancient soils (pointed out) represent time gaps between eruptions. Los Humeros volcano, Mexico.**

within a pumice layer at all geographic locations, even though the thickness of the layer and the overall size of the pumice fragments in it change significantly from place to place. For example, as an individual layer is traced away from a volcano, the pumice fragments within it decrease in size, but the vertical *patterns* within the layer remain exactly the same and so can, with care, be matched up with confidence from one site to another.

Once a field volcanologist has identified each different layer (usually giving each a name to remember it by), its thicknesses and the maximum size of the pumice and rock fragments it contains are carefully recorded at numerous geographic locations. These measurements are then plotted onto a map, and used to map concentric contours of pumice size and layer thicknesses across an entire region. Such 'isopleth' and 'isopach' maps are extremely useful. First, they indicate which volcano the pumice actually came from. Then they can be used to calculate the height of the former eruption column, and thereby estimate the 'mass flux' of the eruption; that is, how much magma was erupted each second. Even the direction and strength of the stratospheric wind at the time of the eruption can be calculated, because wind blows against the stratospheric umbrella cloud, and so skews the thickness and grain size contours into the downwind direction. The stronger the wind, the more elongate the contour patterns become. Finally, the volume of each eruption can be calculated, using the layer thickness contours (isopachs) to estimate the total volume of the pumice in the layer: this can be estimated even if the map is incomplete (as is commonly the case) because pumice fallout layers tend to thin exponentially with distance from the source volcano.

*When* each eruption happened may be deduced from historic records, from buried archaeological remains, or from radiometric methods, which use the regular decay rate of naturally occurring radioactive isotopes in the ash to compute the duration of time since the eruption. The volcanologist thereby painstakingly assembles an account of how the volcano has behaved in recent times, how large and frequent the explosive eruptions have been, the areas affected, and so on. This can help develop a prognosis about what may happen in the future.

## Andesitic explosivity

Magmas that contain rather less silica are less polymerized and so allow gases to escape more easily. One type, common in the Andes, is *andesite*. This is highly explosive, but can also be erupted more slowly, as degassing and congealing sticky magma rises to slowly extrude out of the volcano. Too viscous to flow away as a liquid, the andesite lava instead accumulates at the top of the volcano as a steep-sided, nearly solid hot mass or lava dome, which grows slowly from week to week. As it degasses, the andesite becomes brittle and breaks into a network of cracks that contribute to the dome's instability and aid the escape of gas. Angular blocks soon break off and tumble down the sides of the lava dome, forming a skirt of loose scree.

Andesite eruptions are prone to intermittent violent explosions when gas pressures within the conduit abruptly force out the congealed blockage, suddenly lowering the pressure. The explosions are typically short-lived (Figure 8), and generate shock waves and a spectacular bulbous ash plume that ascends convectively several kilometres above the volcano summit. Fridge-sized blocks of hot lava are ejected hundreds of metres into the air and can be seen following ballistic trajectories picked out by white vapour trails. Such cannon-like explosions are known as 'vulcanian' after 19th-century eruptions witnessed on the Italian island of Vulcano. Luckily, volcanologists are getting better at forecasting when vulcanian explosions might occur, following years of careful monitoring at andesitic volcanoes like Soufrière Hills volcano on the little Caribbean island of Montserrat. This is important for local inhabitants, because periodic increases in dome-growth activity are often accompanied by particularly

**8. A small, short-duration explosion at Fuego stratocone, Guatemala, in 2019 with ephemeral pyroclastic density currents on upper slopes.**

hazardous explosions and catastrophic collapses of the unstable lava dome, generating destructive ground-hugging density currents made of hot gases, lava blocks, and ash that surge down valleys, incinerating and burying anything that lies in their path.

## Basaltic explosivity

Basaltic magmas contain less silica, and more magnesium, calcium, and iron than silicic magmas and are consequently less polymerized and far runnier. This means that dissolved gases can readily diffuse through them, reaching existing bubbles and causing them to grow. The result is far fewer, but larger gas bubbles, and these are able to rise through the more fluidal magma, just as gas bubbles rise gently through a glass of sparkling wine. The bubbles rise up the volcanic conduit, growing and coalescing, and as they encounter lower and lower pressures they grow even larger, eventually bursting out at the surface, releasing the volcanic gases into the atmosphere, whilst squirting out a spray of incandescent, partly congealed magma and ash, in the form of a turbulent fountain that may reach a few hundred metres in height. Much gas is released, but the rising gas bubbles drag out little of the surrounding hot magma along with them, so less magma erupts in a given time. As it is the flux of magma out of the volcano that drives the convection of eruption columns to high altitudes, basaltic explosive eruptions are less cataclysmic: with a lower rate of magma supply, there is insufficient thermal supply to drive a high eruption column. The eruptions are relatively benign—the sort you could watch from a reasonable distance in relative safety. The eruption plume doesn't reach the stratosphere, and can be seen curving away downwind from the summit of the volcano as it is deflected by even modest winds.

The chilled fragments of magma don't get far from the volcano. Instead, they accumulate close to source in a large conical heap, similar to a pile of grain beneath a hopper on a farm. If the eruption continues for a few months the cone may reach more than a couple of hundred metres high. When the sides steepen to 34° the grains tumble downslope, rattling and vibrating as they do so. The faster the tumbling mass of grains roll, the more vigorously they vibrate against each other and the tumbling mass expands, just as molecules in a gas vibrate more vigorously on heating, causing the gas to expand. Smaller grains can then drop through the expanded voids between the tumbling and vibrating grains, whereas larger grains rise to the top of the tumbling mass and therefore tend to roll farther downslope. Consequently, the grains that make up this type of volcano are remarkably well size-segregated and organized into thousands of stacked lens-shaped layers, each pinching out across a few metres, and recording a little downslope avalanche.

A fragment of chilled basaltic magma from this type of eruption is quite different to the pumice of a Plinian eruption: the bubbles are larger—clearly visible without a microscope. The thicker bubble walls mean that the rock looks darker, and is denser than pumice.

**9. Scoria with clearly visible bubbles from a basalt lava, Outer Mongolia. Scale is in cm.**



Such bubbly rock is called 'scoria' (Figure 9). It is not a microfoam and will not float on water as pumice will. Some scoria surfaces are smooth and contorted, reflecting the fluidal nature of the basalt magma as it squirted and twisted though the air (e.g. 'ribbon bombs'). Larger, twisted pieces have pointed, broken ends where an airborne squirt of liquid stretched and necked before snapping. The resultant fragments thin to a point at each end, and are known as 'spindle bombs'. The shapes indicate that the magma had frozen by the time the fragments impacted the ground, the result of cooling by air during flight. Other fragments remain fluidal right to the point of impact, forming a Monty Pythonesque splat on landing; they are delightfully named 'cowpat bombs' or the onomatopoeic 'spatter'.

Eruptions of this type are 'Strombolian', named after Stromboli volcano in Italy's Tyrrhenian Sea. And because the resultant volcanoes are cone-shaped they are called Strombolian scoria cones. Stromboli has been erupting for thousands of years, and has been one of the great spectacles of the Mediterranean since classical times. A three-hour walk to the top (best not undertaken in the midday sun!) is rewarded by a close-at-hand explosion every ten minutes or so. The ground shakes, a puff of black smoke rises, then the sound arrives. At night it is one of the best firework displays, with incandescent bombs of congealing basalt tracing ballistic trajectories through the night sky to land, still glowing, upon the crater walls, where some roll back inside, to be ejected once more. But don't stand downwind, as noxious sulphurous gases are released. This poses something of a paradox: gases have been released pretty constantly over thousands of years but relatively little magma has come out of the volcano, so where does all the gas come from? One idea is that new magma rises under the volcano, releasing the gas, but a large component of this magma is injected into the volcano without making it out of the top. If this is true, the volcano must be getting larger from within. Not surprisingly then, part of the volcano collapses every so often into the sea, causing local tsunami that run up nearby coasts. Every few years some magma does leak out from a vent on the volcano's flank, and it creeps downslope, forming black ribbons and fans of lava. Where the slope is too steep, the heavy lava flow over-extends itself and pulls itself apart into massive, container-sized chunks that bounce and clatter down the talus slope, eventually plunging into the cool sea far below with a great deal of fizzing, popping, and steaming.

Yet people live on the island of Stromboli in relative safety. Although the eruption is explosive, the explosions are relatively small and harmless, with little fallout onto the villages just a couple of kilometres distant. Volcanologists call these eruptions 'cone-forming' to distinguish them from 'sheet-forming' Plinian eruptions. Although the scoria deposited reaches a great thickness, this thins rapidly away from the volcano. Fallout layers from Plinian eruptions may not reach such thicknesses, but their thicknesses diminish from the volcano much more gradually, over kilometres rather than tens of metres, forming a sheet-like layer of pumice that drapes an entire region. The total volume of such a sheet of pumice is far greater than the volume of a cone of scoria, and this reflects the far greater dispersal of Plinian eruptions, and the greater extent of the effects.

At oceanic volcanoes like Hawaii and Galápagos, the basaltic magma is so fluidal it forms incandescent Hawaiian lava fountains. They are generally less than a kilometre high, and magma droplets in them chill in the air, and freeze, forming droplet-shaped glass particles known in the trade as 'achneliths'. These can be found a kilometre or more downwind and are highly characteristic of this style of eruption. Also characteristic are thin, drawn-out filaments of basaltic glass stretched into golden fibres whilst still viscous and airborne; some may remain connected to a dark droplet-shaped glass tear. The filaments, which can reach to more than a metre in length, readily blow downwind, coming to rest in trees and fences, and are known as Pele's hair after the long-haired Hawaiian goddess of volcanoes, fire, and dance. Closer to source, still-hot basalt may land and congeal as steep-sided towers and ramparts of welded spatter. Because the hot fragments stick together, they can form steep structures without collapsing. More fluidal drops and splashes of incandescent magma may, on landing, coalesce back into a continuous liquid lake or pond of lava which, when it overflows the crater, may drain away several kilometres downslope from the eruption as lava flows.

Hawaiian fountains also produce an amazingly delicate rock called 'reticulite' that consists of a honey-coloured sparkling lattice of fine glass filaments. It is as much as 90 per cent air and, although lighter than pumice, doesn't float because each bubble wall has popped, rendering the rock highly permeable.
A remarkable feature is that the bubbles were all of very similar size before they popped. On Hawaii, reticulite forms only at very high (over 1 kilometre) fountains, but exactly how it forms

remains something of a mystery. One idea is that it involves particularly rapid bubble nucleation and expansion within the rising fountain, bursting of low-viscosity bubble walls, and almost immediate chilling in air to preserve its delicate, lace-like structure. Copious volcanic gases are released from Hawaiian eruptions and blow downwind. In Hawaii, weather reports include a vog (volcanic smog) report. Asthma or allergy sufferers in Honolulu, more than 300 kilometres distant, are advised to remain indoors on voggy days.

We have seen how physical properties of different types of magma, including viscosity and gas diffusion rates, result in a wide range of explosive eruption styles, from Plinian, Vulcanian, Strombolian, and Hawaiian. What all these types of eruption have in common is that they are driven by the escape of gases dissolved in the magma.



## Ground-hugging pyroclastic currents

The rise and stratospheric spreading of a Plinian eruption column, with fallout of pumice and ash across a region, is truly awesome and hazardous, but commonly it does not represent the climax of a giant volcanic eruption. It is often followed by an even more devastating phase—of pyroclastic density currents. Being so far from common experience, a pyroclastic density current can be difficult to envisage: a vast, turbulent, ground-hugging cloud of searing-hot pumice, ash, and gas tears across the landscape, destroying everything in its path. It may move as fast as a train, commonly at about 100 km/hr (~60 mph), though sometimes much faster, and in just a few minutes buries everything that it does not incinerate. Nothing survives. Whereas Plinian pumice layers reach a thickness of a few metres, pyroclastic currents can leave deposits tens and even hundreds of metres thick. They can bury entire towns (famously Akrotiri, Pompeii, and Herculaneum) and forests, and fill entire valleys, leaving a landscape unrecognizable.

How do these currents form? During a Plinian eruption, the nozzle (known as a vent), from which the exploding mixture emerges from the volcano, is prone to erosion by the supersonic passage of material. It can crack and rapidly enlarge, so that the diameter of the emerging Plinian jet increases. This means that proportionally less of the erupting mixture is in direct contact with the atmosphere. This decreases the efficiency with which the jet can entrain and mix in cool atmospheric air: when insufficient cool air is sucked into the explosively volcanic jet, the erupting mixture (of gas, ash, and pumice) fails to expand sufficiently to loft buoyantly through the atmosphere: instead it simply collapses straight back to the ground. This is a continuous process, just like a fountain. On impacting the ground the mixture of hot particles and gas is deflected and spreads rapidly outwards. This is a pyroclastic density current—it is denser than the atmosphere, and gravity causes it to spread radially across the landscape (Figure 10).



A pyroclastic density current may travel a kilometre or 100 kilometres from the volcano. How far depends on the slope and how rapidly material is supplied to it by the volcano: a larger mass of material supplied in a given period of time increases the transport distance. As the current passes, it sediments out ash and pumice, gradually forming a thick layer of deposit, known as an 'ignimbrite'. Vast sheets of ignimbrite lie in many parts of the world—Chile, Ethiopia, New Zealand, Turkey, Mexico, Japan, Yemen, and Indonesia to name just a few—testament to cataclysmic density-current eruptions in times past. Most people who dwell upon and work this fertile land have little idea what awesome events produced the vast ignimbrite plains on which they live. Ignimbrite sheets vary from less than a centimetre thick to hundreds of metres in thickness. The thickness depends first on how rapid the deposition is (this in turn depends on how abruptly the current decelerates, for example due to a change in the slope of the land) and, second, on the duration of deposition: pyroclastic currents from the largest eruptions continue flowing for several hours, or possibly days, during which great thicknesses of ash and

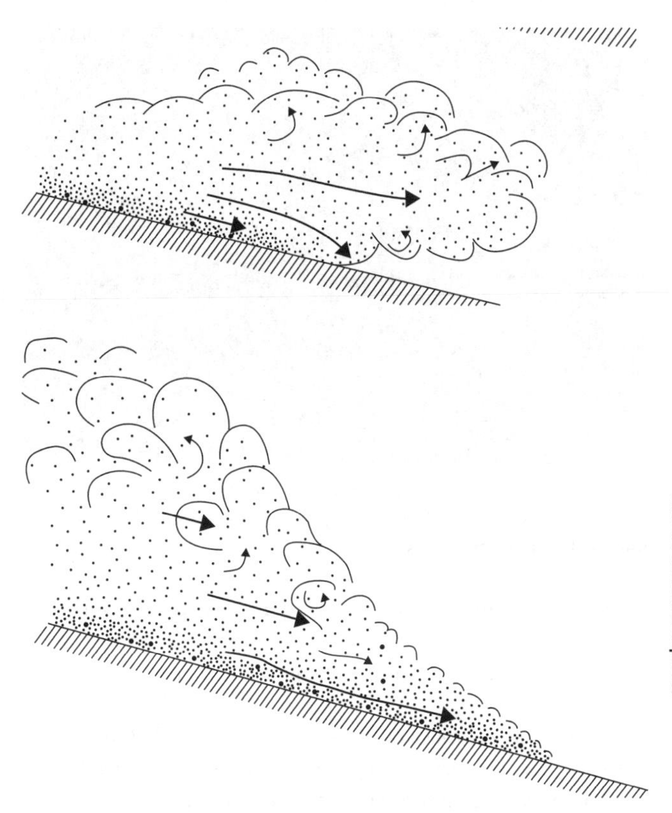

**10. Pyroclastic density currents carry ash and rock fragments rapidly across a landscape. Their concentration is greatest near the base. The layers deposited may record where the dilute component arrived at a location first (above) or the more concentrated component (below).**



pumice accumulate, filling entire valleys during a single event (Figure 11).

At Mount Pinatubo in the Philippines, ignimbrite as much as 200 metres thick filled valleys radiating from the volcano. Despite being subjected to torrential tropical rainfall, this ash was, two years after the eruption had ceased, still at 500°C. Charcoalified

**11. Ignimbrite: the deposit of a ground-hugging pyroclastic density current. Friable pumice fragments (pale, rounded) were abraded during transport. South-west Tenerife.**



remains of the rainforest are preserved in the ignimbrite, and are used for fuel by locals. Yet the ash, as with many ignimbrites across the world, still remains loose and non-compacted.

Elsewhere, some pyroclastic density currents are so hot that the particles they carry are actually viscous droplets of magma. In this case the current resembles a flowing, incandescent turbulent liquid–gas spray. The searing-hot droplets adhere to the ground across which the current travels, and coalesce to form an extensive sheet of volcanic glass. Such currents literally enamel the landscape. Forty-five thousand years ago an entire island in the Mediterranean was enamelled in green incandescent glass: on Pantelleria, a quiet retreat favoured by celebrities such as Sting and Madonna, the glass can today be seen fused to all slopes, valleys, and hills on the island, and even to the vertiginous sea cliffs. This used to be thought of as an exotic, rare volcanic phenomenon; but it is becoming increasingly apparent that in various parts of the world entire regions have been coated in sterilizing sheets of hot glass, with numerous examples in Ethiopia, Kenya, India, Namibia, south-east Brazil, eastern Argentina, southern Australia, Texas and Idaho, USA. In the western USA, individual ignimbrite sheets exceed 100 metres thick, cover several million acres, and have fused together so completely that they are rock-solid and virtually indistinguishable from lavas. What is particularly intriguing about these deposits (known as 'rheomorphic tuffs') is that, even after being transported in pyroclastic currents for many tens of kilometres from the volcano, the particles remained sufficiently hot to congeal together, and so then continued to flow, like an incandescent gloopy oil-slick, wrapping and folding themselves as they did so, like the contorting skin on poured hot custard—though here the folds can be over 10 metres in height. The deposits were as hot as 1000°C and they actually baked the ground across which they travelled, firing soils into vivid-orange terracotta. Break off the lowermost layer of glassy ignimbrite, and the imprints of desert sage can still be seen. These astonishing events are now being reconstructed by volcanologists, but remain unknown by the population at large. Such eruptions have not been witnessed historically, but on a much smaller scale, as recently as 1875, extremely hot density currents left rheomorphic tuffs at Askja volcano in central Iceland.



How much ash and pumice a density current can transport depends partly upon its velocity. So, where a current decelerates, such as where it fans out across a plain, or it encounters a gentler slope, some of the pumice and ash is deposited out. Such deposition will itself change the topography of the landscape across which the density current flows and this, in turn, affects the flow directions of the current, so that the current shifts laterally over time, twisting in response to the gradual accumulation of deposit at certain sites. For example, a valley may become filled with deposit half-way through a sustained eruption, so that later parts of the pyroclastic current are able, for the first time, to surmount a high ridge that had, just minutes before, formed a barrier. The density current then abruptly sweeps down a different, pristine valley that had hitherto remained unaffected by the eruption, causing devastation of yet another river catchment. Such behaviour means that great care must be taken when trying to predict or model where a pyroclastic current may flow (and destroy) based upon an area's topography at the start of an eruption.

## Phoenix plumes

Meanwhile, the pyroclastic current is becoming diluted along its upper surface by the mixing-in, heating, and thermal expansion of cool atmospheric air. With continued deposition, and air ingestion, there will come a point on the ground, say a few tens of kilometres distant from the volcano, where the current carries so little pumice and rock fragments that it actually becomes buoyant. At this point the current simply lofts high into the atmosphere, forming an ash-laden 'phoenix plume', so named because it rises from the hot ash, rather than directly from the volcano itself as in a Plinian plume. Phoenix plumes differ from Plinian eruption plumes also in that they contain mostly fine ash. This is because the ascent velocity of a phoenix plume is insufficient to carry larger fragments of pumice, rock, and crystals, which tend to be left behind and deposited in the ignimbrite.

Wind can blow a phoenix plume considerable distances. Within the plume, moisture causes the ash particles to stick together, and tiny, soft ash pellets rain to the ground, leaving thin layers of ash pellets (Figure 12) upon fields, pavements, and car roofs. These layers of ash pellets provide important clues that an ignimbrite-forming eruption happened somewhere nearby, upwind. But they are only a few millimetres thick and easily removed by wind or rain. The delicate ash pellets disaggregate easily, readily squashing between fingers. Yet, remarkably, some billion-year-old examples are still preserved, enabling geologists to deduce that similar processes have occurred on Earth right back to deep prehistoric times, well before life had colonized the land.

**12. These tiny, soft ash pellets fell onto the ground high above a moist phoenix plume that lofted from a pyroclastic density current (scale in centimetres).**

# Chapter 3
# Volcanoes and water

The popular image of a volcano is a conical hill smoking beneath a blue sky. Yet many eruptions happen during rain, under clouds, snow and ice, or in water. By far the majority of volcanoes on the planet lie beneath the sea—and water can profoundly affect the way a volcano behaves.

## Subterranean magma meets water

When magma ascends through the Earth's crust, it commonly encounters water even before reaching the land surface. This is because rock at depth, beneath the water table, is commonly saturated, as one can see when a well is dug. Water resides in cracks and pore spaces within the rock. Some rock types, like limestone, sandstone, and fractured basalt lava, make superb natural aquifers, which is to say that they allow large volumes of water to pass through them. When rising hot magma encounters groundwater, as in a water-saturated sandstone, the magma may never make it up to the surface to form a lava. Instead, the pore water in the sandstone next to the magma flashes to steam, and the force of the expanding steam within the rock tears the sand grains apart from each other, and streams them away as it escapes. Sand grains at the contact are fluidized: this is when gas is streamed up through a mass of dry sand such that the sand expands, loses its strength, slops about, and can be poured like a fluid.

The hot magma cannot touch the solid wet rock directly because an intervening screen of expanding steam and fluidized sand grains develops immediately at the contact. As soon as the steam streams away, more is generated within the sandstone now next to the magma, causing this, too, to disaggregate into loose grains and stream away, and so on. In this way the solid sandstone is incrementally eaten away by the fluidizing layer of steam, and the magma can flow passively into the space once occupied by solid sandstone bedrock. Bizarrely, the bedrock is able to exert little resistance to the advancing magma because the magma is never really in direct contact with solid rock. The magma cannot press directly onto, or deform, the rock because it only experiences steam ahead of it. Ultimately, large volumes of rock become replaced with magma. Because the fluidized sand behaves as a fluid, it stirs and mingles with the liquid magma, forming swirling marble-like patterns. In this stirred zone, magma is chilled by the (less hot) steam, and it freezes, contracts, and shatters into a myriad tiny angular glass chips.



All this activity happens underground, hidden from our view; but evidence of these subterranean events can be found millions of years later, when the ground surface has been uplifted and naturally eroded away by streams, rivers, and glaciers, exposing for the first time what was once deeply hidden beneath the volcano. For example, former steam bubbles can still be seen preserved within vestiges of the disaggregated, pulverized bedrock along the contact with what was once hot magma. Cracks and bubbles in the now solidified magma are filled with (once fluidized) sand or mud. Clouds of sharp splinters of quenched glassy magma are interleaved and swirled together with the homogenized sandstone, forming marbled patterns. This rock-type is called 'peperite' due to a passing resemblance to coarsely ground pepper.

Magma bodies emplaced underground in this way can be large—traceable for many kilometres—and they form stacked sheets that can exceed a kilometre in total thickness. All this tells us what goes on beneath our feet. Clearly, a large proportion of rising magma never makes it to the surface of the Earth, but just flows sideways under the ground, beneath the water table, for kilometre after kilometre. Not long ago, many such bubbly peperite intrusions across the world were thought to have been lava flows, but we now know this to be incorrect. A source of confusion has been that the bedrock right next to the invaded hot magma shows no sign of having been baked by the magma. This is because the rock was thermally insulated from the hot magma by the ever-present thin envelope of steam, and the wet rock behind could not easily exceed water boiling temperature. Peperites are astonishingly common in ancient volcanic regions, from Chile to South Africa. In Britain they are abundant within the coalfields of central Scotland, at Glencoe volcano in the Scottish Highlands, and in the English Lake District, Snowdonia, and Pembrokeshire

**13. Peperite formed beneath an underwater volcano. Fingers of hot rhyolite magma (pale) froze whilst invading wet, sea-floor mud (black), mingling and shattering. Ynys Dewi (Ramsey Island), South Wales.**

(Figure 13). They are probably forming quietly underground somewhere right now.

Peperites tell us that water–magma explosions don't always burst out at the land surface. The gases (including steam) may escape quietly through permeable rocks to the surface. Large volumes of volcanic gases can be released quietly into the atmosphere through the ground, even when there is no volcanic eruption. But if the gases, particularly steam, cannot escape, the pressures may build up, leading to rupture and a violent explosive eruption.

## Exploding groundwater

Explosive eruptions driven by expanding steam are known as 'phreatic' (from the Greek for well, or spring). They are common and can be highly destructive, the expanding steam sending fragments of pulverized rock several kilometres into the atmosphere, generating density currents and shrouding the landscape in grey dust. Phreatic explosions are particularly difficult to forecast, and famously startled tourists in 1924 on Kilauea volcano, Hawaii, and tragically in 2014 in Japan when tourists, trekking high on Ontake-san volcano to see the spectacular views and autumn colours, were confronted with clouds of rock fragments speeding through the air, killing sixty-three people. More ominously, phreatic explosions can be precursors of much larger volcanic eruptions, where the cause of the groundwater explosivity is ascending hot magma. Not always, though. Phreatic explosions can be caused by sudden pressure release: underground water can be highly pressurized and remain liquid well over 100°C, but if the pressure is released (e.g. as a crack forms in the rock), it can abruptly flash to steam, blasting out the rock and excavating a large crater. Hot groundwater circulates underneath most active and dormant volcanoes, and manifests itself variously as fumaroles (hot gaseous emissions), hot springs, and geysers, in which water and steam periodically fountain into the air to the delight of tourists who gather daily for the occasion. In addition to steam, fumaroles commonly release other volcanic gases, such as carbon dioxide, hydrogen chloride, and the malodorous hydrogen sulphide. The fumes can be toxic, and emissions may persist for decades, while the circulating steam and gases cause chemical reactions in the host rocks, degrading some to clays, as evident in the bubbling, gloopy mud pools of Rotorua, New Zealand, and Yellowstone. In the process, steam-rich gases can cause underground precipitation of a wide variety of minerals, including valuable ones such as gold, silver, copper, and other metals.

## Phreatomagmatic explosions

In some circumstances, hot magma invades an aquifer, causing explosive blasts that rupture the ground surface and eject tiny shreds of hot magma together with smashed-up rock. This particularly violent style of volcanic explosivity is known as 'Taalian' after Volcano Island on Lake Taal, of the Philippines, where such an eruption was witnessed in 1965. The explosions happened when new magma encountered water within the old basalt rock of the island. The eruption devastated the island, and the pyroclastic density currents wreaked havoc among fishing communities along the shores of Lake Taal. Taalian eruptions may not be as large as the Plinian eruptions described in Chapter 2, but they are thought to be particularly violent because of the additional effect of explosively expanding steam. This pulverizes the magma to fine ash, and blasts out a wide, circular crater (Figure 14). Some time after the eruption, a lake may form in the crater, such as the tranquil beauty spots in the Eifel region of Germany, west of the Rhine. Taalian eruption craters are called 'maars' and are found from Alaska to Australia. A ring of tuff (hardened ash deposits), aptly called a 'tuff ring' (Figure 15), commonly forms a low rampart around the crater. The tuff is generally fine grained, but contains large, angular blocks of rock blasted from the aquifer. On the Pacific island of Oahu, tuff rings

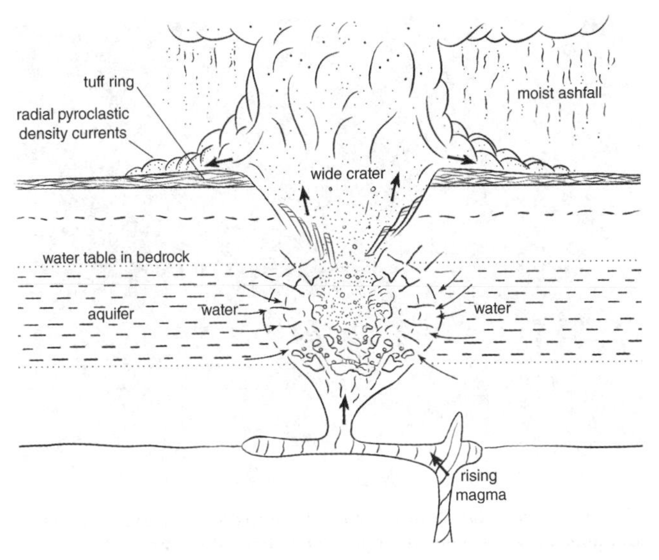


**14. A Taalian eruption, in which rising hot magma encounters water in bedrock, explodes repeatedly, and blasts out a wide central crater or 'maar' with a surrounding low tuff ring.**

**15. The typical wide and low profile of a Taalian tuff ring, where rising hot magma exploded with groundwater in an aquifer. Montaña Pelada, Tenerife.**

on the coastal fringe near Honolulu developed on ancient coral reefs. Rising hot magma encountered groundwater within the fossil reef, blasting out fist-sized chunks of white, fossil coral like cannon balls: these can still be seen where they impacted down into wet, black volcanic tephra.

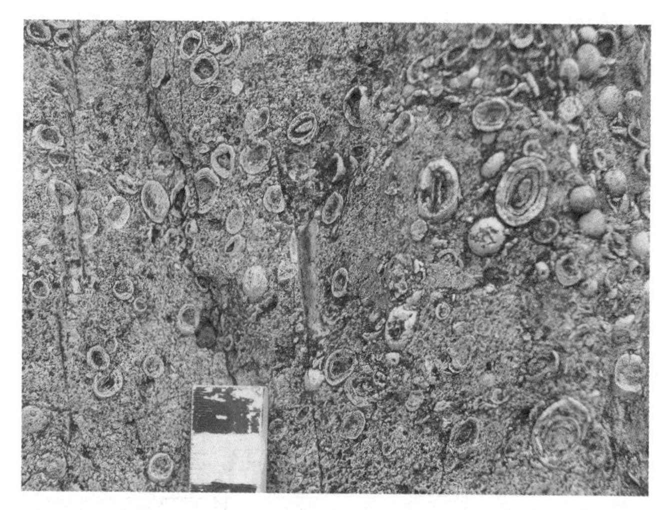

**16. Accretionary lapilli with their characteristic concentric structure and some broken fragments supported in fine tuff: an ancient pyroclastic density current deposit from a phreatomagmatic eruption. Langdale, England.**



Pyroclastic density currents from phreatomagmatic tuff rings leave well-layered deposits that contain pea-sized accretional lapilli (Figure 16) that form when moist pellets (Figure 12) drop from atmospheric plumes into upper, turbulent parts of the density current, where they circulate, and grow additional layers of fine ash.

## Submarine eruptions: pillows and hyaloclastites

Where hot, bubbling and fragmenting basaltic magma fountains beneath the ocean, steam generation is limited by the intense pressure of the deep water, which suppresses boiling. So, by and large, the deeper the ocean, the less explosive the eruption. However, deposits similar to those of Hawaiian fire fountains on land have been found in ancient marine deposits at Heerbornseelbach in central Germany, and accumulations of scoria have been recovered from 2 km-deep ocean floor south of Japan. Such evidence suggests that explosive fountaining may sometimes occur beneath our oceans—even where the high water pressure suppresses the explosive expansion of steam. Such fountains may be driven by the pressure of rising buoyant magma and by the release of gases that were dissolved in it.

But mostly underneath the deep ocean, lavas are produced. Basalt lava underlies all the world's oceans, much of it formed at mid-ocean ridges. The hot melt accumulates within shallower magma reservoirs, which have the form of sub-horizontal sheets of crystal-rich mush. Basalt magma rises from the reservoirs along vertical cracks, known as dykes, towards the seabed, where it erupts and flows away downslope as lava. The eruptions are accompanied by release of hydrothermal fluids (hot water) into the ocean. The fluids contain minerals that precipitate out, forming billowing clouds and gradually building up vertical chimneys to heights of as much as 50 metres before they collapse to form heaps of broken debris. These 'black smokers' may be as hot as 400°C; less hot examples are called 'white smokers'. Submarine eruptions and associated hydrothermal activity have probably gone on since the advent of plate tectonics, billions of years ago, and the fluids released into the water have cumulatively affected the composition of the Earth's oceans.

Depending on the water depth, submarine lava flows may be sufficiently hot to vaporize seawater at the surface of contact, resulting in a thin envelope of steam between the lava and the seawater. As soon as the steam bubbles away, more is instantly generated so that the steam envelope is maintained. This insulates the water from the hot lava, and so seawater even a short distance from the lava is not heated as much as one might expect. This is known as the Leidenfrost effect, after Johann Leidenfrost who first described it back in 1756 in his delightfully entitled 'A tract about some qualities of common water'. It can be demonstrated by letting water drop, first onto a 100°C hot skillet so that the drop of water evaporates away rapidly. But then drop water onto a much hotter skillet and it doesn't immediately evaporate, but skitters around like a tiny hovercraft on a layer of steam. The skittering water droplets survive longer because they are insulated from the hot skillet by a steam film constantly replenished at the contact. It is this insulating effect that enables scuba divers to dive close to submarine lava flows without being scalded.

Basalt lava flows advance underwater by exuding multiple 'toes' or tubes of incandescent lava that protrude into the water and rapidly darken as they chill against the steam envelope. The lava crusts over in seconds, but the brittle crust soon ruptures and another incandescent toe of lava protrudes forward, again rapidly darkening and chilling. This mesmerizing process happens again and again. The lava 'toes' may be 10–60 centimetres in diameter, depending on how fluidal the lava is. Early workers who mostly viewed ancient examples in cross-section thought that these shapes represented accumulations of detached pillow-shaped bodies, leading to the term 'pillow lavas' (Figure 17). This term is still widely used even though it is now recognized that they have the three-dimensional form of interconnected branching tubes.

Molten lava flows down the tube or 'pillow' to supply the advancing protrusion. If it drains away faster than it arrives, a hollow section of tube crust is left and starts to cool. The Leidenfrost effect decreases allowing cold seawater to gain direct access to the tube, rapidly cooling it, causing the gas inside to contract so that the pillow implodes, sending shock waves through the water—to the distress of inquisitive scuba divers nearby. Implosions of this kind shatter the outer crust of the pillows into angular, glassy fragmental debris. When hot magma of any composition is abruptly quenched (chilled), it cracks and disintegrates as it rapidly contracts. The effect is rather like casting a slowly oven-heated wine glass into a bucket of cold water (don't try this!)—the glass shrinks suddenly, setting up internal stresses that cause the glass to shatter into tiny sharp glass shards,

**17. Tectonically tilted pillow lavas on a late Precambrian ocean floor. Porthdinllaen, North Wales. Each 'pillow' is a really a long sausage-like tube of basalt, up to half a metre in diameter.**

known as 'hyaloclasts'. Deposits of hyaloclasts, 'hyaloclastites', are commonly found associated with pillow lavas, and this association is an excellent indicator that the volcanism took place under water. With time, the basalt lavas and associated hyaloclastites that cover the oceans become buried beneath layers of fine ooze that accumulate very slowly from the sunken remains of microscopic plankton, from desert-derived atmospheric dust, and even meteoritic dust.

Submarine volcanoes can build up to form a mound or large seamount. And if the volcanic activity persists so that the top of the seamount eventually encroaches shallow water, the lower hydrostatic pressures allow volcanic gases to escape from the magma more readily and expand, causing the magma to rise more vigorously, bubble, and fragment. Experiments show that quench-induced shattering is particularly effective if the glass contains bubbles.



At shallower depths, steam generated by the interaction of fragmenting magma and seawater bubbles up through the water. This may cause passing ships to sink: ships are designed to float on water with its normal specific gravity, and not upon bubbling water, which has a significantly lower effective density. The tragic loss of the schooner *Island Queen* in the Caribbean Sea north of Grenada in 1944, with sixty-seven people on board, has been linked, speculatively, with sailing over the submerged, bubbling Kick-'Em-Jenny volcano. More recently, in 2011, intense gas bubbling just off the coast of El Hierro in the Canary Islands marked a volcanic eruption where volcanic cones built up on the seabed to heights of over 100 m over a few years, though these never reached the sea surface.

## Emerging islands

As the top of a growing underwater volcano shoals, it becomes increasingly explosive. The magma rips apart more violently to

**18. A Surtseyan eruption in the Pacific at Hunga-Tonga in 2009.**



form smaller ash particles, which are squirted vigorously through the water column. The eruption may continue for several weeks or months. Watery slurries of tephra accumulate on the seabed and sometimes back into the vent, to be erupted again and again.

Once the water depth has decreased to a few metres, the watery jets of tephra, volcanic gases, and steam break the sea surface, and can be seen from nearby ships as dark, tephra-laden fountains accompanied by white billowing clouds of steam. These repeatedly squirt high into the air, and fountain down in arcuate trajectories reminiscent of a cock's tail or cypress tree (Figure 18). As the ash builds up, the volcano itself eventually emerges from the sea, and a new island is born, such as the island of Surtsey, which formed off the south coast of Iceland in 1963, and gave its name to this particular style of emergent explosive activity. The eruption began at a depth of 130 metres and the island emerged within a couple of weeks. 'Surtseyan' volcanoes are much more violent than their on-land (Strombolian) equivalents because the explosivity is enhanced by the explosive expansion of steam.

More recently in 2015–17, a pristine island nearly 2 square kilometres in size, known as Hunga Tonga Hunga Ha'apai, emerged from the south-western Pacific Ocean during a spectacular Surtseyan eruption. A key feature of this style of vulcanicity is eruption through unconsolidated watery ash, so that the conduit has no rigid walls and the sloppy material constantly founders back into the erupting mixture, with which it is re-entrained, reheated, and re-ejected as partly mixed slurries of water, ash, and steam (Figure 19). The fine-grained ash produced is composed of tiny angular shards of steam- and water-chilled glass. These contrast with the more bubbly glass shards that form when frothing magma is shredded during Strombolian or Plinian eruptions in air.



Surtseyan volcanic activity is 'hydromagmatic' in the sense that both water and magma are essential ingredients, but the style is

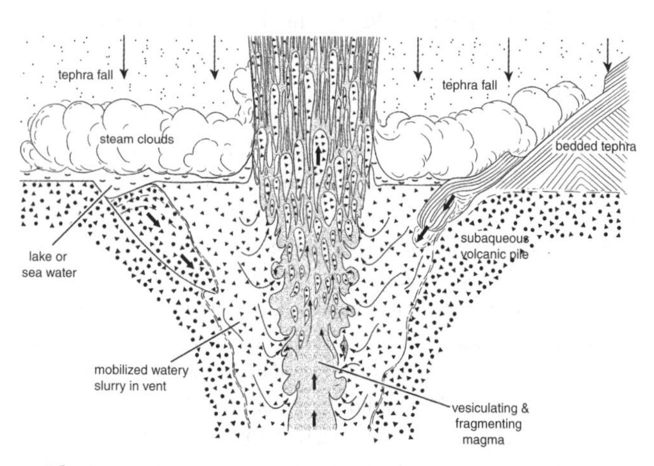


**19. The Surtseyan eruption mechanism in which uprushing magma admixes with a watery slurry in short bursts.**

distinct from Taalian activity, which blasts through solid bedrock and ejects large rock fragments. Surtseyan tuff rings are waterlogged, soft, and highly unstable, particularly prone to collapse and to erosion by waves. However, both Surtseyan and Taalian eruptions are highly pulsatory and produce distinctly layered ash deposits, each layer recording an individual explosion. The steam condenses to water in the atmosphere and rains down onto the loose accumulations of tephra, rapidly gullying them. If the eruptions persist, a Surtseyan tuff ring may ultimately dry out, the glassy ash chemically reacting with the warm seawater causing it to set into harder tuff. Once seawater is excluded, the explosions may become less violent, Strombolian, and a cone of bubbly scoria fragments starts to grow, sometimes accompanied by gentle effusion of basalt lava flows. Lavas, being strong, help stabilize an island, so that it lasts longer before being eroded away by the sea. The initially barren, pristine island soon becomes colonized by plants and animals.

Surtseyan eruptions also occur in lakes. The resultant tuff rings often long outlast the lake itself, and testify to the former presence of lake water in areas that are now desert-dry. For example, they can be seen in desert flats in Oregon's central basin, USA, in central Mexico, and even on Mars, recording former watery conditions. Surtseyan eruptions are known to have happened back to at least 1.8 billion years ago, as preserved in ancient rocks around Hudson Bay in Canada.



## The most violent eruptions

The eruptions outlined earlier are small-scale, involving fluidal basalt. But at some volcanoes larger volumes of viscous, silica-rich magma interact with cold water. Such eruptions would be catastrophic even without the water, because gas escaping from silicic magma tends to rip the magma upwards explosively, as in Plinian eruptions. But from the early 1970s volcanologists began to recognize that some tephra layers were unusually fine grained and extensive, being traced for tens of kilometres with little change in their appearance. The erupting magma appears to have fragmented with such extraordinary violence that it was pulverized to fine particles, and distributed widely in giant atmospheric eruption columns. Intriguingly, even closer to the volcano these ash layers didn't get much coarser-grained and, typically, the source areas were flooded. Volcanologists deduced from this that sea or lake water somehow mixed with the already frothing hot magma, with steam explosivity providing additional force to further fragment the normally walnut-sized fragments of pumice into tiny shards.

Such turbo-charged eruptions were termed 'phreatoplinian', and are widely regarded as the most violent on Earth. The most recent occurred at Askja in Iceland in 1875 and lasted about seventeen hours. It began with a highly explosive Plinian phase, soon followed by ground shaking and a change to a violent phreatoplinian phase, presumably because cracking of the ground caused the vent to relocate to a pre-existing lake at the volcano. The resultant ashfall layer extends from Iceland across Scandinavia and into Poland, but even near the source, 99 per cent of the ash particles are less than 1 mm in size. The eruption column was about 26 km high.

The largest known phreatoplinian eruption in recent times happened about 26,000 years ago, when more than 500 km$^3$ of rhyolitic ash erupted through prehistoric Lake Huka in North Island, New Zealand. We know about this 'Oruanui' eruption from ignimbrites (thick ash layers, deposited from pyroclastic density currents) and a complex succession of ashfall layers some of which extend more than 1000 km across the south-west Pacific Ocean. Some of the ashfall came from eruption columns that rose above the volcano's vent and some came from buoyant phoenix plumes that rose convectively from the ground-hugging density currents. During the eruption, the atmosphere around North Island would have been choked in clouds of moist, airborne volcanic ash.

Tiny ash particles in the moisture-laden atmosphere adhered together forming clusters of ash that grew into pellets, the size of millet seeds. Such pellets fall out through the atmosphere onto the land surface (Figure 12) where they accumulate as white, topography-draping layers. The layers are thin and may not look impressive, but once their true extent is appreciated it can be seen that the volume they represent is huge.

Similar ash deposits in the Philippines indicate that about five giant 'hydrovolcanic' explosions erupted through the 25 km diameter Taal lake during the last 100,000 years. The lake used to be connected to the sea and was navigable by Spanish galleons, but then became disconnected during a relatively small eruption in 1753 and now contains many endemic species newly adapted to fresh water, including the world's rarest sea snakes. Around the lake, cities have been built upon ash layers from the previous large hydrovolcanic eruptions. There have been many small explosive 'Taalian' eruptions in recent times, the last one in 2020, but should another eruption occur on the scale of the five giant prehistoric eruptions, as many as 20 million inhabitants could be at risk. Happily, none is forecast—currently, at least.

The largest phreatoplinian eruptions of all time were far bigger even than this (Figure 20). A phreatoplinian ashfall layer in the English Lake District erupted from Scafell supervolcano, some 450 million years ago. It, too, contains millions of little ash pellets formed by the clumping together of ash within vast, moist atmospheric clouds, and there is evidence that rainfall accompanied the eruption, in the form of fossil raindrop imprints seen on the ash surfaces and tiny erosional gullies that were cut into some ash layers, only to be draped by the succeeding thin blankets of ash. Subsidence and catastrophic ground shaking also accompanied the eruption, literally shaking the strata apart. By comparison, the combined ashfall layers of New Zealand's Oruanui eruption do not exceed 4 metres in thickness, whereas the remarkable ashfall layer from this Lake District 'Whorneyside'

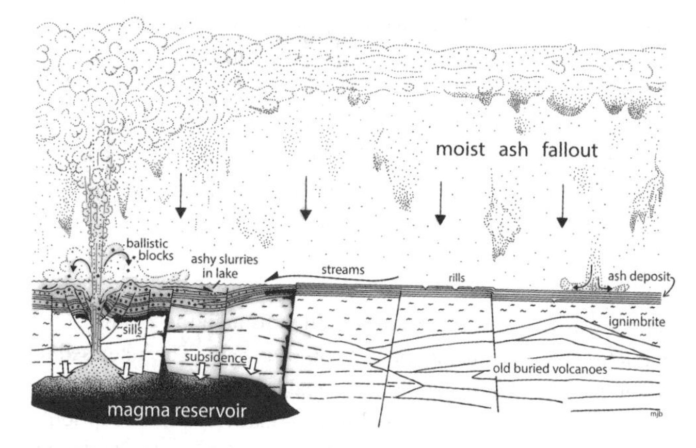


**20. The immense Whorneyside phreatoplinian eruption at Scafell volcano reconstructed, English Lake District.**



eruption is over 25 metres thick. This would not be unusual, except that this thickness is maintained consistently up to 30 kilometres from source, suggesting that the layer was originally very extensive, and therefore of unusually large volume.

As one walks along this thick layer of fine ash northwards across the Lakeland fells from the Duddon Valley, and across the Coniston and Langdale Fells, rare impacted rock fragments within it become progressively larger and larger. The largest are cannonball-sized and these are found in the ash layer farther north, in Seathwaite, Borrowdale, which therefore must have been close to the volcano source. Follow the ash layer north from here, and one starts to see within it for the first time the appearance of fossilized wave ripple, mud-cracks, and other evidence for a shallow lakeshore. Below the fossilized watery lake bed, hot magma can be seen to have leaked into the wet tephra forming abundant, steaming peperite intrusions. Some of these broke surface, forming heaps of blocks and tephra slurries that sloughed away across the lake bed as debris flows, some containing 2 ton blocks of andesite. The walk northwards across the Lake District

National Park therefore takes you from the prehistoric desert land surface, right into the lake through which this particularly violent eruption blasted—a unique experience unmatched at any modern volcano.

There is still much to learn about the physics of these enormous 'phreatoplinian' eruptions. It is not yet understood how tens to hundreds of cubic kilometres of magma can be brought together sufficiently rapidly with similar quantities of lake or seawater to maintain a vast explosive eruption. Perhaps the volcanic jet actively sucks the lake water in like a giant venturi pump—the fast-moving fluid having the lower pressure to draw the adjacent water inwards. In this case the explosive steam expansion would happen within an already-formed Plinian-like volcanic jet, enhancing its vigour. Steam condenses as the atmospheric eruption plume rises, like a giant thunderhead, creating a local weather system in which eruption-triggered rainfall helps flush the fine ash out of the atmosphere (fortunate, because breathing in fine airborne ash is hazardous).



## Watery slurries

An explosive eruption blankets the landscape with loose ash. The ash coats hills, forests, crops, and houses. Rain washes it into gullies, gutters, sewers, and watercourses, clogging them all. Flooding inevitably results—but not ordinary floods, because as the soft stream banks collapse into the streams, so much ash is added that sparkling, tumbling streams transform into dense, swift-flowing ash-charged slurries that ominously lack the whitewater spray normally associated with rapids. Such flows are known as 'lahars' and, unfamiliar though this term is to European ears, it is well known and much feared in volcanic regions of the Philippines, Indonesia, and South America. The destructive power of lahars is immense, readily sweeping away forests, houses, and bridges. During storms and typhoons, lahars strip away ash and loose ignimbrite from higher regions around the volcano and carry it at breakneck speed down river valleys, inundating floodplains and broad agricultural plains. The best response is to evacuate, and this is best achieved through careful advanced planning, with early warning systems and well-rehearsed lahar evacuation drills.

Lahars can vary from freezing to scalding. They dump out thick layers of ashy sediment across the plains and can bury an entire town overnight; afterwards, the location of major roads may be made out only by the remnant lines of projecting telegraph poles. Lahar sediments dam rivers to form new lakes, which tend to grow alarmingly and ultimately burst through after a few months, ripping out bridges, roads, and cables far downstream.

The amount of ashy sediment in a lahar can vary. A large ash content makes a flood 'hyperconcentrated', and its greater density makes it barrel along faster than a clearwater flood, forming standing waves. Sediment is dumped out as the lahar passes. Rapidly accumulating, it diverts the flow obliquely as hidden sand bars develop and shift. When more ashy sediment collapses into a lahar, the lahar 'bulks up' into a debris flow, which is a thicker, rapidly flowing slurry, commonly likened to flowing concrete mix. Debris within this type of lahar may include pumice, ash, ripped-up soils, lava blocks, trees, cars, and wellington boots. Boulders the size of industrial containers are rolled and dragged along intermittently through the lahar in spate, the density of the lahar providing buoyancy, and traction to move them.



## Ice and snow

Many volcanoes grow where there is snow or ice cover, from small glaciers to thick continental ice sheets, such as in Antarctica. Even in hot regions, volcanoes can grow so high that snow and glaciers develop on their upper slopes. Well-known examples include the glacier-draped Mount Rainier that towers above Seattle in the Pacific Northwest, and central Mexico's Pico de Orizaba or

Citlaltépetl, which, at over 5½ kilometres, is the highest volcano in North America and supports nine glaciers.

Eruptions can melt such ice, triggering lahars down the side of the volcano and across the plains beyond. In 1991, Gaucho ranchers in southern Chile saw a lahar advancing toward them like a wall of liquid down the broad Huemules valley, slamming into their wooden cabins and squirting in upon frightened residents through windows and cracks between planks, almost filling the cabins. Survivors cowered within the rafters, and when they later emerged, they were met with an unfamiliar landscape transformed by a 2 metre thick layer of quicksand out of which projected 'bayonet trees', southern beeches bent over in the direction of flow, sharpened to points and stripped of leaves and bark. House-sized megaliths of glacier ice stood upon the flat waterlogged sand. They had been carried 45 kilometres from Hudson volcano where the eruption had melted and fragmented the glacier. Some of the ice blocks were buried within the accumulating sand, and melting during succeeding months left a landscape pockmarked with circular subsidence pits, each surrounded by gaping curved crevasses.

The outlandish landscape would be familiar to Icelanders, where extensive coastal plains have formed by successive sediment-laden glacial outburst floods. Where volcanoes are buried beneath an ice-sheet, the thick ice exerts a confining pressure that initially suppresses explosivity, causing the first volcanic eruptions to be more gentle. Lavas intrude the ice, melting it, and forming ice-bound cavities full of water. Melting also can be caused by hot water and steam from the volcano. The overlying ice subsides, initially forming a circular structure known as an 'ice cauldron' (Figure 21). Eruptions in ice can be similar to submarine eruptions, with the formation of pillow lavas and glassy, fragmental hyaloclastites. These build up, melting progressively shallower levels in the ice sheet, which begins to sag, then cracks above the subglacial volcano, ultimately melting all the way

**21. 'Ice cauldrons with concentric crevasses form by subsidence where a subglacial volcano causes melting. This example in April 2006 at East Skaftár, Iceland, had dropped almost 160 m.**



through, so that the volcano finds itself in a new-formed ice-bound lake. As with a volcanic island emerging out of the sea, the glacial volcano becomes progressively more explosive and violent as it shallows. Eventually it emerges from the lake and dry lava advances over the earlier wet accumulations. Such glacial volcanoes are known as 'tuyas' and characteristically have steep sides where they were buttressed by the ice, and flat lava tops. Their distinctive shape (Figure 22) and internal characteristics are readily identified in many parts of the world, and indicate the presence of long-vanished ice sheets. Because their age can be determined from radioactive isotopes within the volcanic rocks, their height may be used to indicate the thicknesses of the former ice sheet at that location and time. In this way, volcanologists have been able to reconstruct the shape of prehistoric ice sheets, for example across swathes of Iceland, eastern Russia, British Columbia, northern Oregon, and Antarctica. Such studies contribute to our understanding of global climate change.

**22. A steep-sided, flat-topped tuya formed by volcanic eruption into a former thick ice sheet. Herðubreið volcano, Iceland.**



Ice-confined meltwater builds up within the ice sheet and is prone to catastrophic escape when the ice is breached. It then bursts out as spectacular floods known as jökulhlaups. These race to the sea, sweeping up and redistributing great volumes of sediment across braided coastal *sandur* plains. Jökulhlaups accompanied recent eruptions from Katla, Grímsvötn, and Eyjafjallajökull volcanoes in Iceland, washing away roads and bridges, making the coastal plains impassable.

# Chapter 4
# Lava

Most magma remains underground and eventually solidifies to form a crystalline rock. The evidence for this is the abundant igneous intrusions of all shapes and sizes found within the Earth's crust, such as the granites that form the Torres del Paine and Yosemite, the gabbros that form the magnificent black Cuillin of Skye and Rum in Scotland, and the countless basalt dykes that lie parallel to each other beneath the floors of our oceans. Some magma, however, finds its way up to the surface, particularly when it is buoyant relative to surrounding denser rock.

Within a kilometre or two of the ground surface, the rise of hot magma is accelerated by the growth of gas bubbles, which increase the magma's buoyancy. If the bubbles remain in the magma as it rises towards the Earth's surface, they multiply and cause the magma to froth, as in the explosive eruptions we have described. But if the gases escape from the magma soon after they come out of solution, for example by leaking out along fractures, or by diffusing away into surrounding permeable bedrock, the magma may ascend more slowly without being ripped apart, and it eventually emerges from the volcano as lava. As the lava flows, bubbles in it variously grow, stretch, coalesce, or collapse, and some become frozen-in as the lava chills and solidifies (Figure 9). These frozen bubbles are known as vesicles, and their various shapes and sizes provide the volcanologist with useful information about how and when the lava degassed, flowed, and cooled. Later, minerals may grow inside the vesicles from hot fluids, sometimes producing spectacular banded agates. The rising and degassing of magma can also cause new crystals to grow within it. The interacting effects of crystal and bubble growth are increasingly the subject of study by laboratory experimentation using apparatus that can gradually decompress or deform hot home-made magma, while the bubbles and crystals growing within it are monitored by CT-scanners, similar to those used in medicine, that use X-rays to see the magma and growing bubbles in three dimensions.

## Effusion

Some lavas are runny, and flow rapidly in thin sheets and streams. Rare lime-rich lavas called carbonatites are so runny that they form miniature dark, silvery lava flows as little as one centimetre thick, that flow rather like water, yet that are as hot as 500–600°C and glow in the dark, as can be seen by those sufficiently intrepid to trek up into the crater of Ol Doinyo Lengai volcano in Tanzania.



But most lavas are several orders of magnitude more viscous than water, and ooze along as thicker lava flows. This is because the flow of a lava is hindered by large chains of silicate polymer. On Kilauea, Hawaii, until recently you could watch basalt lava flows about 8 centimetres thick, advancing as little lobes and 'toes' 20–50 centimetres long. Each emerging toe-like lava protrusion was initially incandescent but darkened in seconds as its surface chilled against the air (Figure 23). The forming skin around the new toe immediately started thickening to a black crust, which then snapped open, and a new little incandescent lobe of lava emerged and rapidly chilled to a crust, and the process repeated. Over a few days, many thousands of advancing and chilling toes resulted in the overall advance of the lava. For many years the rate of advance was often not great and, with care, one could stand by these lavas and be entertained by the endlessly fascinating

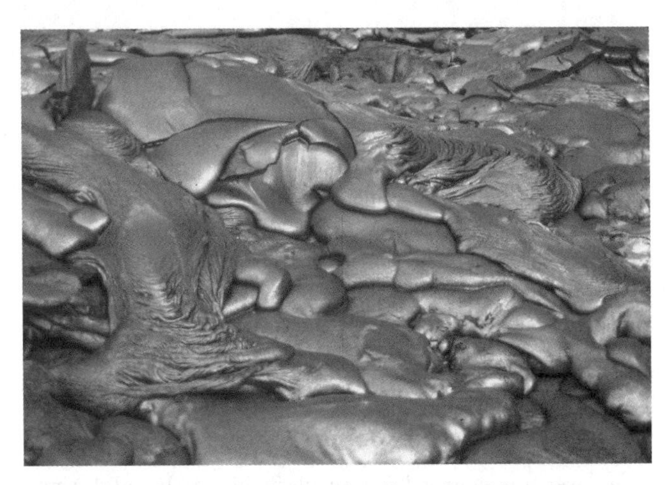

**23. Pahoehoe lava toes budding and cooling as basalt lava advances. Kilauea volcano, Hawaii.**



budding, advance, and chilling of new toes. This type of lava was called 'pahoehoe' by indigenous Hawaiians.

There are many forms of pahoehoe, including the well-known ropey variety that forms when the surface skin wrinkles and shears, 'toothpaste lava' with its parallel grooves, and slabby and spiny varieties of pahoehoe. The top, glassy surface glistens like metal and is a delicate filo pastry of fine bubble-rich layers that crackle and scrunch like eggshells when walked upon. Where pahoehoe flows advance down steeper slopes, the glistening toes and tubes of lava resemble the spilt entrails of some giant butchered animal, so that this variety of pahoehoe takes the gory term 'entrail lava' (Figure 24).

So long as they are given a wide berth, such lava flows are not particularly hazardous to human life. But they are persistent and unrelenting, and will destroy any forest, house, or town in their path: prompt evacuations are required when they encroach upon

**24. Glistening entrail pahoehoe lava that flowed over a steep scarp on Kilauea volcano, Hawaii.**



developed areas, and fatalities can arise from, for example, asphyxiation, exploding fuel, and disease in displaced populations, as at Goma (Nyiragongo volcano, Congo) in 2002. Basalt lava flows may sometimes advance as rapidly as a speeding car, but usually only during peak phases within eruptions that overall may persist for as long as decades, gradually covering up more and more ground.

But this is a far cry from the tsunami-like torrents of lava that used to be envisaged inundating entire landscapes in minutes. Confusion stemmed from the great thickness of many ancient lavas, such as those covering vast tracts of the Russian steppes, the Deccan region of India, Paraña in Brazil, east Greenland, the Inner Hebrides of Scotland, and the north-west USA (Figure 25). In these 'large igneous provinces', some individual basalt lavas reach as much as 100 metres thick and clearly travelled tens to hundreds of kilometres from source. Scientists had proposed that in order to travel so far, they must have flooded across the

**25. Four thick basalt lavas, once thought to originate from catastrophic floods of liquid basalt but now thought to have crusted over and inflated to their present thicknesses. Columbia River, USA.**



landscape extremely rapidly—any slower and their progress would soon be checked by chilling and solidification. Incandescent deluges of biblical proportion were envisaged, with flow-fronts tens of metres high, towering over the hapless dinosaurs and other doomed creatures scattering in the path of these 'flood basalts'.

Ideas began to change when volcanologists watched modern pahoehoe lavas flowing at Kilauea volcano in Hawaii. The leading edge of the lava field sometimes was only a few centimetres thick and advanced rather slowly, as described earlier by the emergence of toe-like breakouts (Figure 23). But simultaneously something else was happening, at a rate so slow that it had been overlooked entirely. Fluidal basalt flowing beneath the thickening upper crust of the lava was slowly jacking up the crust, so that the lava actually inflated like a waterbed—its top gradually rising by several metres over a period of days. Once the upper solid crust of the lava has cooled for a week or so, it can support the weight of a person standing on it, even while searing-hot lava continues to flow a few metres underfoot (best not try this—it's easy to lose a foot should the crust be thinner and break). It's likely you would not even notice that you are rising as the lava beneath you imperceptibly inflates.

It is a humbling truth that despite abundant, clear field evidence for lava inflation on display at volcanoes across the world, field geologists didn't really appreciate its significance until the advent of time-lapse photography. In hindsight, one can see evidence for inflation in many old basalt lavas worldwide, in the form of multiple layers of sheared bubbles in thick upper crusts of pahoehoe lavas, each bubbly layer frozen-in as the crust thickened downwards. Mounds, known as 'tumuli', are found in groups on the tops of lavas. They have the form of gigantic blisters, with open cracks radiating from their summits, separating outward-tilted lava crusts. The cracking, lifting, and tilting of the upper crust of the lava evidently occurred due to the pressure of hot liquid basalt still flowing within. Late-stage toes of glistening lava can sometimes be seen having oozed out of the cracks. Then there is the famed MacCulloch's tree standing over 12 metres high, buried in lava on the Isle of Mull in Scotland. It was discovered two centuries ago and has for generations attracted hikers along the Ardmeanach peninsula. It was once part of a coniferous forest that met its end engulfed by lava 60 million years ago. A 40 metre thick flood of dense lava would have toppled this tree effortlessly, yet it evidently remained standing in its life position, because the original lava *flow* that encroached it would have been just centimetres thick, prior to inflating. Although the basalt lava that entombs it is now 40 metres thick, this is merely the (stationary) lava, not the thickness of the (moving) flow that gave rise to it.

Where lava flows down a steep slope it tends not to inflate and so remains thin. But where the base of the slope flattens out the lava may inflate to many times its original thickness, supplied from upslope by a head of more liquid lava between the formed crusts.

The key to long-distance flow lies in the insulating properties of the thick upper crust: once developed, the crust slows the cooling and degassing of the flowing liquid beneath, and enables the basalt lava to remain molten for longer and thereby travel so far from source—incandescent torrents as once envisaged are not required. Nevertheless, the name 'flood basalt' lives on for the leviathans of the lava world.

Basalts are the most common type of lava on Earth, covering the ocean floor. Beneath water they initially advance in a rather similar way to that of pahoehoe on land: small entrail-like toes or tubes protrude forward, rapidly cool, crack, and squeeze out the next momentarily incandescent 'pillow' or tubular protrusion. As a result, many ocean floors are covered by pillows and attendant hyaloclast fragments. Marine geologists have long distinguished between these and so-called 'sheet lavas', which are several metres thick and extend for long distances. At least some sheet lavas are simply subaqueous varieties of pahoehoe, in that their initial advance was by protrusion of a series of pahoehoe toe-like pillows before crusting over and slowly inflating by injection of more hot fluid lava within.

Back on dry land, pahoehoe flowing downslope may move faster than its viscosity can accommodate. It then tears itself apart like hot, sticky toffee with elongated connecting strands that thin and snap. The resultant bubbly shreds of scoria have fractal, irregular shapes with thousands of sharp protrusions on all scales. During flow, these serrated fragments are shoved and jostled forward as scrunching piles of jagged, spiny rubble. Deep in the interior, the lava is hotter and still contains dissolved gases, and so remains fluidal, and this drives the flow forward, invading the enveloping mass of rubble, and sometimes protrudes out of the top as irregular spines several metres high, sending rubble cascading down on all sides. Hawaiians call this rubbly type of lava a'a (Figure 26).

**26. A'a lava with characteristic spiny and rubbly surfaces, Etna.**

Pahoehoe can change into a'a, but the transformation is irreversible—degassed and chilled a'a will not coalesce back into pahoehoe. An a'a flow may be a metre to several metres thick, with a steep leading edge of rubble that tumbles and rolls forwards, bulldozed by the incandescent interior. It advances slowly but inexorably across fields and roads, incinerating and demolishing buildings while the local inhabitants look on askance. How far it flows depends on its rate of production, how fluidal it is, the duration of eruption, and the gradient of the slope. In 2018, pahoehoe and rubbly a'a lava flows overwhelmed residential housing estates on Kilauea volcano, Hawaii. A'a has engulfed ski resorts on Etna, and part of the fishing town of Heimay in Iceland. The advance cannot be prevented, and attempts to deflect the flow path to an adjacent area, using restraining barriers, have limited success.

Insuring a house downslope of recent lava activity on basalt volcanoes can be difficult, even though one cannot be sure whether the next lava to flow across a particular site will occur within the next ten years or thousand years. The hazard posed is chiefly to property and infrastructure, and is best mitigated by good planning. I (MJB) recall once seeing the 1930s painted church of Kalapana 'Star of the Sea' on Hawaii stranded at the side of the road on pallets, in transit, ready to be relocated to a less hazardous location. The rest of the village, however, was abandoned to incineration and burial by the lava.

A'a affects the landscape long after the flow has stopped. It is notoriously difficult to cross by man or beast—it is easy to break an ankle on the loose clinkers, and sharp rock surfaces readily tear the skin. In the American West, a'a badlands were a great place to escape a posse, so long as you were desperate enough to suffer. There is no water because a'a is permeable and any incoming rain soon trickles deep down into the rock below. In such semi-arid landscape, it takes many centuries before the irregular lava wasteland gradually becomes colonized and ultimately develops vegetation and soil.



## Extrusion

Many lavas are not able to flow so freely as pahoehoe and a'a. The polymerized networks in silica-rich magmas, such as andesite and rhyolite, render them viscous and brittle, and they fracture as they move. Some rhyolites are virtually solid as they extrude to the surface, and there is abundant breakage into blocks. Block lava flows (not to be confused with a'a) can be as much as 50 metres thick as they move along, and their upper parts are particularly brittle. Movement of the semi-molten interior causes this upper part to break, creating a thick, overriding carapace of faceted, angular blocks from the size of a brick to a shed. As the lava creeps forward, the upper blocks slowly jostle as they are carried forward on their slow conveyor. There is an apocryphal story of a field geologist settling down for a picnic lunch and enjoying watching the scenery pass by—only to realize he was dining out on a slow-moving lava flow.

As lava blocks reach the front of the flow they tumble down to accumulate in a skirting heap of scree. Standing ahead of such a lava flow, you see this apron of scree incrementally rattling forward, with the occasional block rolling and bouncing ahead. The lava creeps inexorably forwards, bulldozing the splintery blocks ahead, overriding some and shoving others aside. The resultant internal anatomy of a block lava can be best seen in ancient examples, cut through by erosion, for example at Crater Lake in Oregon, or at Wrynose Pass in the English Lake District: an upper zone of loosely packed angular blocks passes down into a folded, flow-banded central zone several metres thick, which represents the once liquid interior, with a basal zone of blocks that had rolled over the snout of the flow to be overridden as the flow advanced like a caterpillar track. Whereas pahoehoe and a'a are associated with basaltic volcanoes, block lavas are more commonly associated with andesitic or rhyolitic continental volcanoes, such as in Italy, New Zealand, and the South American altiplano.



Crystals struggle to grow in block lavas: the various ions needed for their growth diffuse towards them only very slowly through the highly polymerized melt. So it is not unusual for block lavas to solidify without crystallizing at all, resulting in spectacular black splintery volcanic glass, such as the striking obsidian lavas in Lipari (Italy), Tequila (Mexico), Yellowstone, and Newberry volcanoes (Figure 27). In Neolithic cultures obsidian was prized for making tools such as stone axes and arrowheads, and the distinctive chemical compositions of the artefacts have allowed archaeologists to trace the source lavas, and thereby reconstruct sophisticated ancient trade-routes, for example around the Mediterranean region.

Dissolved gases, too, diffuse very slowly, and bubbles fail to enlarge and become trapped so that the dark obsidian locally pales into the rock foam that is pumice. Upper parts of some block lavas and lava domes are made entirely of pumice. Pumiceous lava domes grow in volcanic lakes, where they form islands. House-sized

**27. Block lava, with angular blocks of rhyolite obsidian, Newberry volcano, Oregon, USA.**

chunks of pumice periodically calve off, like icebergs, and float gently away across the lake, ending up stranded on the lake shore as isolated monoliths. These giant pumices can be seen near Guadalajara's Rio Caliente in Mexico, and around the shores of Lake Taupo in New Zealand.

As a viscous lava slowly deforms, any bubbles, cracks, and zones with incipient crystals become streaked out and thinned. In lavas, this stretching happens within the conduit of the volcano and while the lava moves along the land surface by laminar flow. It results in perfectly parallel, millimetre- to centimetre-scale banding, known as flow-banding. As the flow proceeds these bands slowly become twisted and crumpled into spectacular folds, and these folds are themselves progressively tightened, extended, and thinned into long, sheath-like shapes. Included glassy or pumiceous blocks can soften within slower-moving, still hot parts of a block lava, where they, too, become flattened and stretched into lens-shaped streaks and bands. Years later, volcanologists may analyse the intricate internal folding patterns to decipher how the lava moved, and in which direction. On a larger scale, the upper surface of a block lava may crack or crumple, to form concentric folds several metres high (Figure 28).



Most rhyolite lavas do not flow far from source, but some do. In 2011–12, a rhyolite obsidian block lava extruded from Cordón Caulle volcano in southern Chile, and flowed over 3 kilometres from source, aided by the insulating effects of the upper blocky crust. The steep margin of the block lava flow rose 35 metres high, and sounded like breaking plates as it advanced. Every so often, the hot interior would break through the skirting dam of blocks, protrude forward as a new viscous lobe, and cool. But in the Snake River Plain region of Idaho and Nevada, USA, individual rhyolite block lavas with similar features extend for more than 40 kilometres and have volumes in excess of 200 $km^3$. Their stupendous volume and flow distances are thought to have been enabled by a combination of unusually high eruption rates and

**28. Block lavas with short tongue-shapes, and folded broken tops, Montaña Blanca, Tenerife.**



temperatures as hot as 1000°C. However, there is still much to learn about how, and why, they erupted in this way.

## Lava domes and spines

Block lava may simply pile up where it first reaches the surface because it is just too viscous to flow away. Some is virtually solid when it extrudes from the volcano and, rather than flowing, it deforms mainly by cracking, slipping, and grinding. If the eruption continues, the pile of lava grows almost imperceptibly to form a substantial mound, and, eventually, a hill several hundred metres high. This happened in 1944, when, on 1 May, a lava dome appeared from beneath a wheat field near the farming village of Fukuba in Hokkaido, Japan. The dome, called Shōwa-shinzan, gradually grew during the next seventeen months and was faithfully recorded by the local postmaster, Masaeo Mimatsu, who drew successive profiles on to paper attached to his post office window. Nowadays, changing growth rates can be monitored in three dimensions by repeated scanning with pulsed laser beams from the ground or from aircraft, or using successive GPS-positioned satellite images. In 2008, a steep-sided lava dome of rhyolite grew to 120 metres high at Chaitén volcano in southern Chile, accompanied by explosive eruptions and ash emissions, requiring the evacuation of the nearby town.

Gravity acting on the growing mass of a lava dome may cause it to spread outwards, by scrunching along shallow fractures known as thrusts. Meanwhile, angular blocks are shed all around, producing a fringing apron of scree. This was witnessed in the lava dome that grew within the 1980 crater of Mount St Helens in Washington, USA. Volcanologists can recognize the characteristic squat shape of hills and mountains that originally formed in prehistoric times as lava domes. Examples include Montaña Guaza near the port of Los Christianos in the Canary Islands and the immense Lassen Peak in the Cascades range of Northern California, which grew 1200 metres high, about 27,000 years ago. Spectacularly flow-folded interiors and steep blocky margins of ancient rhyolite domes can be seen in cross-section in northern Snowdonia, Wales, and around the summits of Scafell and Seathwaite Fell in the English Lake District.

In rare cases, lava inside a volcanic conduit simply rises intact to form a virtually solid lava 'spine', like a concrete tower being elevated on hydraulic jacks. The diameter of a lava spine reflects the width of the conduit up which the magma is rising. Like stiff toothpaste squeezed out of a tube, the sides of the lava spine show vertical scratches and grooves, formed as the solid lava scrapes past jagged irregularities in the conduit walls. In 1902–3 a lava spine on the Caribbean island of Martinique famously towered 300 metres—the height of the Eiffel Tower—above the crater of Mont Pelée for months before decrepitating and collapsing into a heap of blocks.

## Cracks, explosions, and block-and-ash flows

Volcanologists can listen to the sound of rising viscous magma. Sound waves through the Earth's crust arise from the breaking of the magma. Recorded on seismographs, they can be used to forecast what a volcano may do next. For example, a growth spurt accompanied by elevated noise-levels as the magma repeatedly breaks may herald a phase of violent vulcanian explosions. Rapid growth can destabilize the dome. A dome interior contains pressurized gases, and when confining pressure is abruptly released, say by a rock-slide or avalanche, it can explode with considerable violence.

Tap a block from a collapsed dome lightly with a hammer, and it may miraculously spring apart into a little cloud of lava fragments and ash. This is immensely satisfying and it takes quite a long time before the average geologist can be persuaded to resist tapping every block within eyesight. What is happening here is that the blocks are thermally stressed through cooling-contraction while gas held in microscopic pores is pressurized. Fracturing abruptly releases the stress. The same phenomenon accompanies block collisions within hot rock-avalanches from lava domes, and so the amount of fine rock powder increases dramatically until the avalanches transform into highly mobile 'block-and-ash flows'. These hot flows travel surprisingly rapidly down valleys, carrying hot, bouncing lava blocks the size of a fridge or garage enveloped in dusty gas, burying and burning houses, bridges, and roads. Block-and-ash flows are a particular hazard downslope of growing lava-dome volcanoes, such as during the recent eruptions of Unzen volcano in Kyushu, Japan, and the Soufrière Hills volcano on the Caribbean island of Montserrat. Local residents were displaced for months, some not able to return. Fortunately block-and-ash flows rarely travel for more than a few kilometres. But the lava domes that produce them can continue growing for several years, prolonging the hazard.

## Architectural lavas

As hot lavas cool they contract, and cracks known as cooling joints develop. These are typically arranged perpendicular to the cooling surface, and so a horizontal sheet of lava tends to develop vertical joints, creating a set of vertical, polygonal columns (Figure 29). Columnar joints are common worldwide, and can be viewed on a boat trip to Fingal's Cave on the tiny Isle of Staffa in Scotland, and on the iconic Devil's Tower in Wyoming. The near vertical orientation of the columns at Devil's Tower indicate that this spectacular feature is the eroded remains of an original horizontal layer, rather than a vertical cylinder as in a volcanic conduit.

The width of columns reflects the rate of cooling. Thin columns form where cooling was rapid, whereas thick columns, such as at Devil's Tower, reflect slower cooling. Some thick lavas have a lower level of thick, tall columns—the 'colonnade' (after classical architecture)—with an overlying 'entablature' of smaller, more chaotically arranged columns that formed more rapidly, possibly when cold, surface water penetrated unevenly down into the cooling lava (Figure 29). Spectacular rosettes of radiating columnar joints form where heat was lost radially outwards, as in a filled lava tunnel; or inwards, as when lava envelopes a standing tree and then cools around it (Figure 30).

Vertical columnar joints also form when hot ignimbrites cool, and within tabular-shaped intrusions called sills, where magma became trapped underground, to slowly cool and contract. Andesite lavas, in contrast, tend to develop sub-horizontal 'platy' joints, a few centimetres apart. The reason for this difference remains enigmatic. Cooling rhyolite lavas often develop curvy-planar 'sheeting' joints, spaced a few millimetres apart. Sheeting joints appear to propagate along already formed flow-banding, including around the hinges of previously formed flow folds.

**29. Columnar joints in a basalt lava at Fingal's Cave, Scottish Hebrides. Irregular, smaller columns at top of cliff.**

**30. A basalt rosette or daisy wheel with columnar cooling joints radiating around a former standing tree (central hole). Isle of Mull, Scotland.**

They mimic the swirling patterns of folds formed earlier as the lava was still flowing.

There is much still to learn about many aspects of lava flows. For example, what controls the flux (the rate at which lava issues from a volcano within a given time span)? How do different dissolved gases, and the growth of bubbles and crystals, affect the strength of lava and how easily it flows? What causes those giant, very long hot rhyolite lava flows to erupt, and travel so far? What controls whether magma erupts quietly as a lava flow or catastrophically as an explosive eruption? As laboratory techniques improve, experimentation with realistic materials and conditions, and even using samples of real, hot flowing lava, is increasingly contributing to our understanding.

Lavas are important building blocks for many volcanoes, and each type of lava produces a characteristic landscape. These landscapes can remain barren and impenetrable for centuries: desolate lava terrain is well known on Iceland, 'Craters of the Moon' in Idaho, and Timanfaya in Lanzarote. But with time, the lava weathers and develops clay-rich vegetated soils. Soil formation is favoured by the presence of ashfall layers and by warm moist conditions, as seen on parts of the ocean-islands of Kauai, Vanuatu, and Fiji, and of the Paraná and Deccan regions of Brazil and India.
The startling red colour of some soils between ancient lavas, for example in Scotland, may partly be the result of past warm, moist climates, but in many cases the colour reflects intense baking of the soil by the overlying lava. Large areas of the planet are underlain by lava, but once these are vegetated, flooded, or buried, one needs to look closely to see that a volcanic landscape lies beneath.

# Chapter 5
# Making and breaking volcanoes

What does a volcano look like? It may be a hole in the ground from which magma and gas issue, a snow-clad mountain with a summit crater, or a barely discernible gentle rise across a spread of wheat fields. Or it may have the form of a placid lake-filled basin in the landscape, or a smoking ridge deep beneath the ocean. Volcanoes vary enormously in scale, shape, and longevity. Some erupt just once, others continue erupting intermittently for millions of years. Many collapse, leaving the surrounding landscape littered with a jumble of debris. Over time, volcanoes are relentlessly eroded away by wind, glaciers, and rivers, so that many ultimately are destroyed before they can become preserved within the planet's rock archive of prehistoric events.

## Volcano architecture and construction

The smallest volcanoes are supplied by a single magma-filled crack or dyke, and these tend to erupt just once. Such an eruption may last several hours or months, but eventually any magma remaining in the crack cools and solidifies and the volcano becomes extinct. Such 'monogenetic' volcanoes include Taalian and Surtseyan tuff rings, and Strombolian scoria cones that litter the slopes of many larger volcanoes. It is not unusual for houses to be built on these volcanoes, even in the summit crater, and the residents can enjoy the view with little fear of a repeat eruption.

Once formed, scoria cones endure longer than tuff rings because they are so porous that rainwater drains through them, whereas tuff rings are composed of less permeable, hardened tuff, and so rapidly develop radial rills and gullies as surface water runs off down their slopes.

One type of volcano, a dome field, is an area of scattered and overlapping silica-rich lava domes. It may erupt intermittently, in time building up the irregular landscape of stubby hills, each surrounded by the blocky deposits of rockfall avalanches and block-and-ash flows that crumble away down the steep slopes.

In marked contrast, less viscous basalt lavas can flow several kilometres down gentle slopes, and into broad topographic basins. The basins gradually fill by repeated lava flows, forming extensive lava plains, such as the vast, now-forested Siberian traps that formed a quarter of a billion years ago. Such lava plains may subsequently be elevated by earth movements to form 'plateau basalts'. These typically become incised by glaciers or steep river canyons forming distinctive stepped landscapes, as seen in parts of Iceland, East Greenland, and the Deccan region of India.

Nearer source, successive runny lavas may build up hills that are so subdued that they are easy to pass by without noticing them. They are known as 'shield volcanoes' as each has the form of a round shield lain upon the ground, convex side up. The Snake River Plain of southern Idaho, USA, is formed of a great number of overlapping basalt shield volcanoes with slope gradients of less than 5°. Despite their large size, it takes a while to spot the various gentle 'peaks' of the individual shield volcanoes amongst the seemingly endless sage desert and potato-growing plains (Figure 31). But when one suddenly arrives at the rim of an awesome river canyon cutting deeply through the stark landscape, the vertiginous cliffs reveal a multitude of very gently inclined basalt lavas, and one becomes aware of the seemingly countless successive basalt lava eruptions that, over time, gave rise to this vast and sparsely populated

**31. Easy to miss: one of many low profile basaltic shield volcanoes on the extensive Snake River Plain, Idaho, USA.**



region. Fresh water gushes from hundreds of springs in the basalt and cascades into the canyons. Stored within the highly fractured lavas, the water gives life to the former desert, transforming it, with human help, into a major region of US agricultural production.

Other shield volcanoes have a very different origin. Cataclysmic explosive eruptions send swift and searing-hot pyroclastic density currents in all directions, and these deposit extensive layers of pumice and ash, known as ignimbrites. The ignimbrites first fill and then bury any irregular topography and leave very gentle upper slopes that ultimately taper out several tens of kilometres from source. This gives the volcano its characteristic low, shield-like shape. Ignimbrite shield volcanoes are found within East Africa's rift valley, in New Mexico, and in the Atacama Desert region of northern Chile and Peru. They tell us of repeated and immense volcanic eruptions, each of dramatic scale. Such cataclysmic events are relatively unknown in the popular imagination, because they occur too infrequently to have featured in recent history. The eruptions can be separated by tens of thousands of years, far longer than the collective memory of the human race.

## Collapsing calderas

During a large eruption, so much magma is rapidly expelled from the shallow reservoir beneath the summit region of a shield volcano that the overlying ground is left unsupported. It then collapses dramatically, creating overnight a new topography—a large depression known as a caldera. A caldera formed in this way is distinct from a 'crater' (which is a hole blasted out of the bedrock by an explosion): calderas are larger—anything between one and 100 kilometres in diameter, depending on the volume of magma that came out. As the caldera founders, its steep sides, formed so abruptly, are highly unstable and collapse inwards as a series of landslides; it is not unusual for massive blocks up to a kilometre in size to collapse into the growing caldera. The subsided floor of the caldera also may be intensely fractured, and hot gases escape upwards forming fumaroles, as seen by thousands of tourists annually at Campi Flegrei caldera, at Pozzuoli Bay, Naples, Italy, arguably the most dangerous volcano in Italy.

With its basin-like shape, a caldera soon fills with rainwater, producing a near circular, and sometimes surprisingly deep caldera lake (Figure 32). These make stunning beauty spots such as 'Heaven Lake' on Changbaishan volcano, also known as P'aektu-san, which erupted on the China–Korea border in AD 946, leaving a layer of ignimbrite around the volcano, and showering pumice and ash across 1.5 million square kilometres, as far away as northern Japan and Greenland.

A new volcanic caldera formed in 1991 at Mount Pinatubo in the Philippines. At first it was a brand-new, smoking, 2 kilometre wide hole in the ground surrounded by sheer, teetering cliffs.

**32. This 594 m deep lake surrounded by a steep circular scarp records caldera subsidence during a large explosive eruption. The small volcanic cone (Wizard Island) grew later. Crater Lake, Oregon, USA.**



Rainwater soon began to collect, forming a small acidic pond, which deepened into a circular, acidic lake. A few years later the rising lake levels became cause for concern because, should the lake overtop and breach the rim, ensuing catastrophic lake-burst floods would inundate large areas downstream. An overflow canal was engineered by the Department of Public Works and Highways, to stabilize the lake level. In the years since, the acidity has largely ameliorated and the area is now a serene hiking destination, attracting bathers.

There are places on Earth, known as 'hot-spots', where volcanic activity has persisted intermittently for millions of years. This happens where heat wells up deep beneath the crust (Figure 1), causing partial melting and, over the long term, replenishing the supply of rising basalt magma. Volcanoes at such sites can grow to truly immense dimensions. The Hawaiian, Galápagos, and Canary islands are just the summit tips of huge volcanic superstructures that rise up from ocean depths of up to 6 kilometres, to reach altitudes as high as 4 kilometres above sea level—they are the highest mountains on Earth. As can be seen when viewed from a distance, they are giant ocean-island shield volcanoes. Most are basaltic, but others (Tenerife and Gran Canaria) include ignimbrite shields. Many have large central calderas formed by repeated subsidence events. Some (e.g. Tejeda caldera on Gran Canaria, and Las Cañadas caldera on Tenerife) collapsed during successive violent ignimbrite eruptions, whereas others formed when basalt lava flows drained the subterranean magma reservoir, such as on Fernandina volcano in the Galápagos. Basalt magma readily travels sideways along cracks in the ground and such lavas may issue from a vent quite some distance from the centre of the volcano. Some ocean-island basalt volcanoes have taken several millions of years to grow layer by layer. They have erupted countless times, and are made up of a truly staggering number of radial lavas.



## Giant landslides

Viscous, silica-rich lavas such as andesites do not flow so far, so that a steeper volcano grows over time, producing the archetypal peak known as a 'stratocone'. The elegant Mount Mayon in the Philippines, and the many lofty cones along the Andes, and along the west of central and North America are mostly of this type. Lavas that flow down a stratocone generally have the three-dimensional shape of a long ribbon extending down from the crater. Ribbon-shaped lavas often flow down steep rills eroded into the cone, and on reaching lower slopes they may spread out to form broader lava fans.

Most sediment on Earth accumulates in valleys and topographic hollows. Volcanoes are unusual in being the only large-scale landform that can build skyward by the addition of new material to the top. But this has a consequence: volcanoes can only grow so high before they become unstable and collapse. In the absence of modern structural engineering, Gothic cathedrals of northern

Europe were built ever higher until eventually some collapsed. Volcano construction lacks engineering design, and is also limited by structural failure and collapse.

Before 1980 it was not widely appreciated that tall volcanoes ultimately collapse. But on 18 May 1980 the south sector of Mount St Helens in Washington State, USA, abruptly fell off in a giant landslide. The volcano had grown to stand one and a half kilometres high from its base by successive additions of lava and ash. It had become increasingly unstable. Ultimately, new magma injections within caused it to bulge alarmingly. It then failed catastrophically when shaken by an earthquake, and a giant debris-avalanche sped downslope and across the surrounding landscape, part even overtopping a nearby mountain ridge to be deposited on the far side. A highly astute young field volcanologist named Harry Glicken studied the mass of debris, and mapped out the distribution of the different rock types involved. He noticed that all the rock types had suffered intense fracturing. Solid rock remained in coherent house-sized blocks, but internally these blocks were intensely shattered with a network of irregular fractures, right down to a microscopic scale. Glicken realized that had the landslide moved as a normal rock avalanche, the intensely fractured blocks of rock would have jumbled into smaller rock fragments which would then have separately bounced, rattled, and rolled along, smaller fragments separating out from larger ones as they did so. But this had not happened—the landslide deposit was composed largely of relatively intact, rather squidged chunks of the volcano, so much so that Glicken was able to reconstruct the internal geological architecture of the former volcano by notionally fitting the giant, rather mangled chunks back together again, as with a jigsaw puzzle. So how did the landslide mass move, and what caused the pervasive fracturing?

Before these questions were answered satisfactorily, he and other volcanologists started recognizing the tell-tale effects of colossal volcanic landslides at other volcanoes around the world. Soon each year was greeted by a plethora of scientific publications documenting Mount St Helens-like landslides from stratocones in Japan, Mexico, Alaska, Turkey, Chile, and the Aleutian islands. The giveaway features that allowed such landslides to be inferred were large arcuate collapse scarps high on the volcano, overlooking gigantic spreads of hummocky, irregular deposits stretching out from the base of the volcano (Figure 33), and in which road cuts and quarries revealed the characteristic pervasively micro-fractured and deformed large blocks slid from the former volcano.

The huge landslides that form in this way are technically known as 'debris-avalanches' to distinguish them from other, less catastrophic types of landslide. They are known to happen also in the Himalaya and Alps when melting and retreat of valley glaciers reduce the support for towering mountainsides. It is interesting that some of the non-volcanic examples, such as at Flims in eastern Switzerland, which was, at 43 cubic kilometres, the Alps'

**33. Stretching off to the horizon, the hummocky surface of a giant debris-avalanche deposit formed when Socompa volcano (behind viewer), near the Argentina/Chile border, collapsed abruptly 7200 years ago.**

largest landslide, produced deposits with the same hummocky shape and internally fractured masses of rock as those from collapsed volcanoes, so it seems these debris-avalanches moved in the same way.

Only one other type of deposit on Earth closely resembles them. These form when kilometre-sized asteroids strike, ejecting rock debris upwards and outwards. The resultant layers of mangled rock strewn around the impact crater may, finally, be the clue to how these giant landslides move. If a mass of solid rock is subjected to cataclysmic shock waves, each part of the rock mass is subjected to rapid alternations of high and low pressure. Rock is strong under compression (which is why concrete in construction is commonly 'prestressed' with steel tendons that keep it under compression, and therefore stronger), but much weaker and liable to fracture when under extension (pulled apart). The American impact specialist Jay Melosh suggested that rock masses around both giant landslides and asteroid impact-craters are subjected to particularly intense seismic shock—a rapid succession of pressure waves. When numerous shocks pass through the rock in quick succession, it fractures when in extension and so can move rapidly across the land surface, despite being essentially solid rock. This explains why the deposits of debris-avalanches look so different from those of other types of avalanche, which are looser granular flows of bouncing rock fragments that roll and clatter along, like someone pouring breakfast cereal into a bowl.

The occurrence of a giant landslide will, however, not cease the activity of a volcano. Outpourings of lavas and ash typically continue, leading to the growth of a new edifice, in some cases reaching the height of the former summit of the volcano prior to the collapse. Careful examination of volcanic landslide deposits reveals that they are not a one-off event. Rather, stratocones tend to repeatedly grow and then collapse in giant debris-avalanches, rebuilding and collapsing again and again. So our impression of what constitutes a stratocone has fundamentally shifted: gone is the simple cone-shaped diagram with parallel layers from our textbooks. In its place is a remnant cone with landslide scarps, surrounded by multiple layers of hummocky debris strewn widely around the base, and a younger cone starting to grow. Indeed, after several major collapses (e.g. Colima volcano in Mexico may have collapsed on nine separate occasions), most of the volcano's mass may eventually lie within the extensive apron of landslid material, rather than within the body of the young extant cone.

It is therefore unsurprising that the largest volcanoes on Earth—the giant ocean-island volcanoes of the Hawaii, Canary, Azores, Cape Verde, Society and Austral archipelagos—also collapse when they grow too large. The shape of some of these islands, such as El Hierro (Canary Islands), and Oahu (home of Honolulu), looks as if giant bites have been taken out. Large valley-shaped collapse scarps, some 5 to 10 kilometres wide, extend from the summit to well beneath sea level. Beyond these, extending tens to hundreds of kilometres across the ocean floor, are the characteristically hummocky landslide deposits, revealed to perfection by sonar-imaging from survey ships. Being submarine, the deposits are difficult to study directly; but rare examples have parts preserved on land, such as in the southern desert of Tenerife, where wadi erosion has cut 'barranco' canyons through the deposit of the giant Abona landslide, spectacularly revealing the characteristic hummocky shape and giant rafted blocks with pervasive microfracturing and shearing. Most oceanic landslides are far larger than those at stratocones, with deposits covering tracts of ocean floor several times larger than the area of the island that generated them. Should such an enormous landslide enter the water sufficiently rapidly, the displaced seawater would generate destructive tsunami capable of travelling to distant shores. Tsunami in the Pacific are all too well known, but the number of oceanic-island volcanoes in the Atlantic suggests that a threat of tsunami to Atlantic shores also exists. In practice, however, it is difficult to prove a direct link between a particular volcanic landslide and the record of a resultant tsunami that impacted the opposite side of an ocean. To do this would require meticulous geological fieldwork on the deposits, and a means to establish that both the volcanic landslide and the distant tsunami occurred exactly at the same time. Establishing such a link is a non-trivial challenge because debris-avalanche deposits recycle older rocks rather than creating new ones, like fresh lava that can readily be dated using radioisotope methods.

## Sediment stores

Sedimentation around a volcano in the immediate aftermath of an explosive eruption is convulsive—it chokes and overwhelms pre-existing river drainage systems. Heavy rainfall brings floods—lahars. At first, lahars occur each time there is heavy rainfall, but in the years following the eruption, sedimentation gradually returns to normality, and more stable river courses become established once again.



Where topography causes a lahar to decelerate, the ashy sediment it contains is unceremoniously dumped, forming thick accumulations across the valleys and plains below. A few months or years later, river channels slice through these recently formed 'sediment stores' and mobilize them further down the watercourse, only to be dumped out again at a new location several kilometres farther downstream. In this way, the ash and debris from a large eruption is intermittently shifted, in a series of discrete stages over months, towards the coast. Hence, urban populations, far downstream of a volcano, may become severely impacted for the first time several years after the eruption ceased.

Ultimately, most of the ash and pumice deposited around a volcano ends up washed into the sea, wherein swirling, gravity-powered turbidity currents carry it to greater depths along the ocean floor. Some 5 $km^3$ of ignimbrite erupted from Pinatubo volcano, Philippines, in 1991, had so disappeared within ten years

**34. A temporary sediment store: thick, loose deposits dumped from lahars of the 1991 eruption of Mount Pinatubo are now being eroded by rivers and being carried to the sea (person for scale).**



(Figure 34). The ultimate record of this major explosive eruption now resides as sediment in the bottom of the South China Sea.

All volcanoes that rise high above the landscape are prone to erosion, and the rocks and deposits are eventually washed into the sea. The towering edifices of Mount Fuji in Japan and Mount Rainier, which overlooks Seattle, USA, may look as if they will last forever, but in the immensity of geological time they will ultimately be stripped almost completely away by nature's relentless action (Figure 35). How quickly this happens depends upon the local climate and the volcano's stability. But inexorable erosion means that the best record of past eruptions lies not on the volcano itself, but within the layers of sediment far away, on the sea floor. How to read this record, however, and decipher from the various layers of volcano-derived sand and ash what eruptions happened, their styles and scales, requires a meticulous forensic approach that is even now only an emerging specialist field.

**35. The twin stratocones of Teide (left) and Pico Viejo (right), Tenerife, with Strombolian scoria cones in the foreground.**



The stripping away of volcanoes is illuminating. As layer after layer of a volcano is removed over time by glaciers and rivers, deeper and deeper parts of the volcano are revealed, including older deposits, magma pathways, eruption-induced fractures, magma stores, and a host of other inner working parts that make up a volcano. Normally these are hidden from view and may only be inferred from the evidence of what comes out of the top of the volcano. But where the internal anatomy of a volcano has been exhumed by geological processes, it affords us amazing insights into what makes them tick. Examples of exhumed ancient volcanoes abound, and where these are exposed they reward meticulous attention. In our account of volcanoes, it is this immense resource that we turn to next.

# Chapter 6
# Hidden volcanoes: tales from the past

British volcanologists are so often asked 'but surely there are no volcanoes in Britain?', perhaps seeking assurance that no local cataclysm is likely. But the answer, of course, is yes—there are hundreds of them. There is the submarine volcano of Ramsey Island in west Wales; tuff rings in the NE face of Buchaille Etive Mòr, Glencoe, and at Glaramara in the English Lake District; Llywdd Mawr caldera volcano near Porthmadog in Wales; the volcanic necks of eastern Fife, Scotland; Calton Hill basalt volcano near Buxton in the Peak District; the fluted, valley-filling ignimbrite on the Sgurr of Eigg; the arc volcanoes of Rhobell Fawr (southern Snowdonia), south Shropshire, and Charnwood (Leicestershire); the amazing ring-shaped eruption conduits of Loch Ba (Mull) and Slieve Gullion (Northern Ireland); Eashaness stratocone in Shetland, and the black Cuillin of Rum and Skye (Hebrides), to name just a few. These aren't going to erupt—their activity ceased millions of years ago. But they form the fabric of the landscape, and make some of the more spectacular scenery of the British Isles.

Most countries similarly abound with innumerable ancient volcanoes, from the Oslo region of Norway to outer Mongolia, from the Canadian shield to southern Chile, and from Tasmania to the Sudan. Many have not yet been explored or studied. There are ancient volcanoes in Hungary, Slovakia, Spain, Siberia, Morocco,

Libya, Mexico, across Russia and Italy. In the USA they can be found from New Hampshire to south Texas. Often people living on and around them are oblivious to the fiery origin of the rocks over which they live, work, and commute. Historically, it was the younger and more obvious ancient volcanoes that were first recognized, such as in the Chaîne des Puys of central France.

So what is the point of studying these volcanoes if they are not going to erupt? Well, for a start, they offer five crucial perspectives that would not be clear from the study of active eruptions alone.

First, where volcanoes have been exhumed—uplifted and stripped back by glaciers and rivers—they offer us vital glimpses *inside* a volcano, just as doctors carefully dissect bodies to understand the layout and interconnectivity of internal organs and the complex networks of blood vessels and nervous and lymphatic systems. Many medical advances would not have been made had dissection not been an option, with understanding gleaned only from superficial symptoms. Take a look at the immaculate geological field maps of the glacially decapitated volcanoes on the Isle of Mull, Scotland. It is a pleasure to see the intricate patterns of criss-crossing arcuate fractures, dykes, lavas, and sills that allow the shifting history of volcanic activity and the attendant subterranean magmatic events to be elucidated.

Secondly, extinct volcanoes give us the opportunity to find out what really happens when volcanoes erupt within ice-sheets, lakes, deep oceans, or within bedrock. It is not that easy to poke around in such deep, cold, and wet places for detailed study of the processes—particularly when a volcano is letting rip. So, to see these hidden events by directly examining the contacts between ancient lavas with lake beds, or in former subglacial or submarine settings, has proved endlessly fascinating and instructive.

Thirdly, dissected volcanoes can reveal how volcanic processes vary and develop with time; for example, how a single major eruption begins, flares up, matures, and ceases—and also how a new volcano is born, grows, collapses and evolves over hundreds of thousands of years. Sequences of events like this are beautifully recorded in the ancient rock record—but you need to know where to look for them, and then learn the craft of rigorously 'reading' the rocks.

Fourthly, studying older volcanoes has led to new discoveries of awesome styles of eruption, some far larger and hotter, or just quite different, than anything known from observing active eruptions. This should perhaps not be surprising, because the entire human experience is a mere snapshot relative to geological time, and represents far too brief an interval to reveal the full spectrum of volcanic behaviour. If, like a mayfly, you only lived for a day, you would gain little experience of changing seasons, eclipses, comets, music festivals, wars, or cultural shifts. But the rock record usefully presents a viewing window of much longer duration, enabling us to find out about things that just don't happen that frequently.

Finally, researching really ancient volcanoes—on a time-frame of billions of years—provides us with a long, planetary-scale perspective. It is evident from such investigations that volcanism has fundamentally changed. In the distant past the Earth was hotter and the volcanoes on it were different to how they are now, as we will see in this chapter. This is what motivates geologists to reconstruct events from the Earth's ancient rock record—amazing things have happened on our planet, scenarios far more interesting than anything we could have dreamt up. We just need to carefully gather the evidence together in order to reveal them.

Unfortunately, as with many archives, the older the record, the more fragmented and poorly preserved some of the material may be. Some ancient volcanoes have been irretrievably tectonically mangled almost beyond recognition; others are barely exposed fragments and their tales are beyond deciphering. But every so often, searching unearths a rare gem, preserved to perfection. And it is upon these exceptional antiquities that research efforts focus.

Go back in time, and one sees ancient caldera volcanoes that record stupendous eruptions that dwarf the largest eruptions seen in the last 100 years, such as Pinatubo. La Garita in the silver mining district of southern Colorado is an oblong caldera 35 by 75 $km^2$ in size. It records an explosive eruption so large that it is difficult to visualize, with the emplacement of an astonishing volume of 5000 $km^3$ ash and pumice, apparently at a single sitting. It took over thirty years of field study in the San Juan mountains by Peter Lipman and his colleagues to unravel the immense scale of this 28-million-year-old super-eruption, including a succession of several other devastating eruptions that issued from the same site over the subsequent several million years.



One can only speculate what the effects of such a monstrous eruption would be. More recently (a mere 2 million years ago), Yellowstone famously erupted a volume of 2450 $km^3$ during one of three well-known super-eruptions. Yet new research to the west of Yellowstone has led to the discovery of several explosive super-eruptions of comparable and even larger size than this. And these were even hotter—they enamelled entire landscapes with 60 metre-thick sheets of incandescent rhyolite glass, abruptly sterilizing 23,000 $km^2$ tracts around what is now the Snake River Plain. As our planet is further explored, the sites of more super-eruptions are being discovered from Japan to New Zealand, and from the USA to the Bolivian Andes, so it is arguably too early to gauge just how frequently these major events occur on our planet. Luckily for humanity, however, the larger the eruptions, the longer seems to be the interval between them. This pattern explains why there are many historic records of smaller eruptions, yet only *prehistoric* records of the very largest eruptions. It is not that super-eruptions have stopped happening—there is no sign of that—it is just that they happen so infrequently that one hasn't happened in the very recent past and, hopefully, will not happen in the very near future.

## The Welsh marvel

The highest mountain in Wales, Yr Wyddfa, or Snowdon (Figure 36), is part of an ancient caldera volcano 12 kilometres in diameter that, in Ordovician times, lay within shallow seas. Its story was carefully pieced together over many years by the irrepressible Welsh geologist Malcolm Howells and his team of survey geologists, who used an old caravan in the Snowdonia National Park as a field base. A catastrophic eruption produced the 'Lower Rhyolitic Tuff'—an ignimbrite that reaches over half a kilometre in thickness, burying the caldera even as it foundered. The subsidence was asymmetrical, like a giant trapdoor with the hinge in the west. But the trapdoor was not coherent: it comprised a series of foundered plank-shaped segments that each tilted to a different angle, leading to the concept of a 'piano-key' caldera.

**36. The present glacier-sculpted shape of Yr Wyddfa (Snowdon) in north Wales little reflects the original shape of the ancient caldera volcano (see Figure 37) that is revealed in the rocks.**

Glassy rhyolite magma then oozed up through the ignimbrite, forming bulbous lava domes, on which spectacular contorted flow-banding and a carapace of angular blocks is still seen, despite its immense age. Some magma reached the surface to form lava flows, only to be heaved upwards by the still restless caldera below, to be eroded by wave action into sea stacks of lava: the fossil sea stacks still stand halfway up the mountain. The volcano then deflated, and the stacks foundered deep beneath the waves, to be buried by watery accumulations of basaltic scoria erupted from a newly emergent Strombolian volcanic cone (Figure 37). Meticulous fieldwork by Peter Kokelaar on and around the precipitous north face of Snowdon, interrupted by occasional falling gloves and cameras from tourists perched on the summit



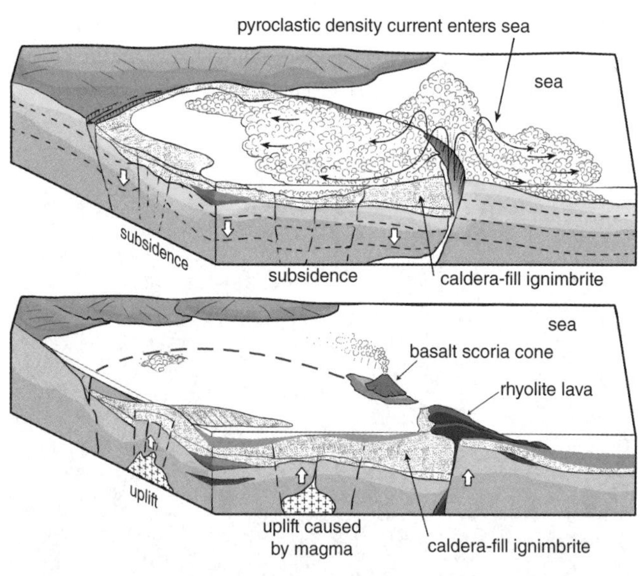


**37. Snowdon caldera foundered beneath shallow seas, 450 million years ago (upper picture), followed by the emergence of small volcanic islands that were successively moved up and down as underground magma shifted (lower picture).**

above, led to his discovery of the 'Britannia vent'—the volcano's eruption conduit magnificently displayed in vertical section. Hot basalt magma later rose up the conduit, then diverted sideways along a crack to form a horizontal sill. In the adjacent cwm, Kokelaar found where basalt lavas oozed out from the sill in the flank of the scoria cone. By this time the Strombolian cone had once again subsided beneath the waves, the sill within it feeding a submarine delta of basaltic pillow lavas. The phenomenal three-dimensional vision of this volcano, coupled with the fourth dimension—time—elegantly demonstrated (using ancient sea level as a reference frame) how restless caldera volcanoes rise and fall, like the heaving chest of a sleeping giant, in this case in and out of the sea. The processes revealed are relevant to modern volcanoes. In historic and recent times there has been comparable rise and fall of the Roman Macellum of Pozzuoli, near Naples.

Subsequent research in the nearby English Lake District National Park led to the discovery of several uplifted caldera volcanoes, including Scafell and Langdale volcanoes. Detailed mapping of Scafell caldera, which took about twenty person-years, showed how its original floor (clearly visible in the glacially dissected landscape) became fragmented into enormous fracture-bound blocks that subsided at different rates and at slightly different times during a brief but spectacular super-eruption. As searing-hot density currents inundated the foundering landscape, huge gashes, 100 metres wide and a kilometre long, opened up in the ground, and rocks avalanched in from growing, unstable fault scarps. Horizontal ground simply tilted, forming steep slopes. The rocks tell a story even more startling than the more fevered imaginings of Hollywood disaster movies.

Different scenarios emerged from comparable studies elsewhere. At Glencoe volcano in Scotland (Figure 38) three distinct explosive eruptions were each accompanied by phases of caldera subsidence, in a more rectangular pattern, controlled by long-lived crustal fractures. With the present-day abundance of bare

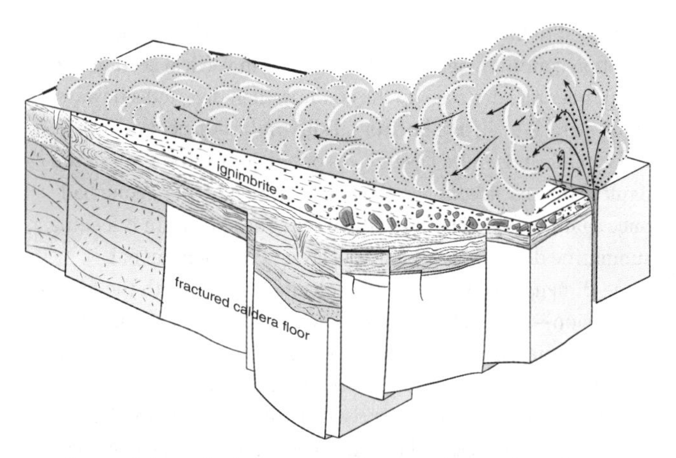


**38. Some of the explosive events at Glencoe caldera volcano in Scotland as carefully reconstructed by volcanologists.**



rock on upper mountain slopes and the deep glacial incision, Scafell and Glencoe are amongst the best-exposed caldera volcanoes known worldwide. Other, equally instructive examples will no doubt be discovered in less accessible regions of the world and will richly reward detailed study.

Other investigation techniques can also be brought into play. Mechanical models of how a caldera collapses are constructed, either using layered sands and clays in a laboratory, or by numerical modelling and computer simulation. Experimental volcanology is a growing and complementary approach to fieldwork with several benefits. In a model one can systematically isolate and change an individual parameter, such as the thickness of a layer of rock, the strength of rock, or the viscosity of a liquid, to help discern what particular factor controls the way things happen. Volcanologists therefore measure spreading rates of syrup on different slopes to learn how lavas flow; they explosively decompress gas-bearing liquids in shock tubes to mimic explosive eruptions, and blow bubbles into viscous liquids to measure foam formation. They compress and shear hot, melted rocks to measure how they respond, and they observe crystals and bubbles growing in hot glasses within laboratory ovens, using three-dimensional X-ray tomography. These all represent early attempts to recreate mountain-scale volcanoes in the laboratory, and they complement the exertions of the field geologist.

## Giant volcanoes of the past

In the time represented by recorded human history, volcanism has ranged from tiny Strombolian cones and gentle lava flows to the awesome explosive eruptions represented by Krakatoa and Mount Pinatubo, in which tens of cubic kilometres of magma are expelled in one mighty paroxysm. These eruptions, from small to large, can generally be fitted easily within the standard context of plate tectonics, with basaltic lavas mostly welling out where tectonic plates continually split apart, and more explosive eruptions being associated with the magmas released above subduction zones. And, overall, volcanic activity of this kind on Earth has generally been more or less evenly spaced.



However, study of ancient strata shows that there have been times in the Earth's past when, during relatively brief intervals, volcanism has flared up briefly and dramatically over wide regions, in a pattern that is more difficult to relate to the more regular mechanism of plate tectonics. These flare-ups are, perhaps unsurprisingly, also the stage for the mightier individual eruptions, such as those of La Garita and Yellowstone.

Among the well-known structures such as mountain belts and tectonic plate boundaries, a simplified geological map of the whole Earth reveals a set of large irregular patches, like an angry rash across the planet. The largest patches are as big as entire countries, and mark regions where tens of *millions* of cubic kilometres of lava and ash poured out on to the landscape within just a few million years. They are called 'large igneous provinces' or LIPs. The most famous examples are the step-like 'trap' terrains of Siberia and the Deccan region of India, where thousands of individual giant lavas rest above one another, but there are notable examples too scattered across both North and South America, in the Afro-Arabian region and southern Madagascar, in Greenland and the high Arctic regions. Some of the largest examples lie mostly underwater, in enormous thickenings of the ocean crust—for instance around the Kerguelen Islands in the southern Indian Ocean, and the Ontong-Java plateau in the south-west Pacific.

Early work on these LIPs focused on the voluminous basaltic lavas that create such stunning landscapes, with some individual lava flow-units estimated to be as much as 10,000 cubic kilometres. More recently, it has become better appreciated that several LIPs also include large volumes of rhyolitic magma, expelled as violent explosive eruptions. For instance, a LIP in North America at which immensely thick basalts erupted around the Columbia River some 17 million years ago also produced numerous rhyolitic super-eruptions, most recently centred on Yellowstone.



What generated these LIPs? They are not related to the edges of tectonic plates that control most of Earth's volcanism. Rather, the magmas leak up right through tectonic plates, whether continental or oceanic. Several are related to hot-spots, which seem to represent places where enormous, slowly rising plumes of hot (solid) mantle cause voluminous melting.

When a mantle plume first impinges upon the base of a tectonic plate it heaves up the landscape above. In response to this uplift rivers tend to cut down, and deep canyons form. Buried within strata beneath the North Sea is a spectacular fossil landscape of kilometre-deep canyons that were incised when such a plume lifted the crust, 55 million years ago. The uplift was followed by a paroxysm of volcanism. It is thought that this is how new oceans are born. In the event, the North Sea failed to crack open sufficiently to form a new ocean. Rather, part of the north Atlantic Ocean began to open and lavas and ashes formed another LIP that now extends from Scotland to Greenland and north-east America.

## Diamond volcanoes

One kind of volcano has connotations for humans quite distant from visions of fiery mayhem, instead symbolizing love and eternal devotion. The mineral diamond is a form of carbon, compressed tightly into a molecular framework harder than any other natural mineral. The crushing pressures needed to make diamonds are only found hundreds of kilometres below the Earth's surface, and to bring these gems from those great depths needs a special kind of volcanism, called kimberlite volcanism. The story of kimberlites holds a fascination and mystery quite in keeping with its precious cargo.



Almost all kimberlites are so ancient that surface traces of the volcanoes have been mostly eroded away. All that is left are volcanic pipes and dikes filled with the smashed remains of dense rock and mineral fragments—including diamonds—from deep in the Earth's mantle. These rocks have been usually altered by subterranean fluids so that much is now a mush of the legendary 'blue clay' sought by generations of diamond prospectors.

Something like 5000 examples of kimberlite have now been found around the Earth, and they have a curious pattern in time and space. They all occur on the continents—none have yet been found in oceanic crust. And, though ancient on human timescales, most are the product of an Earth in middle age. Rare examples do stretch back into rocks up to 3 billion years old, but most are less than a billion years old, clustering into pulses that link with major episodes of continental reorganization. Until recently, the youngest examples known were some 30 million years old, but a few years ago an example, found in Tanzania's Igwisi Hills, was dated to just 12,000 years old. Perhaps another phase of kimberlite volcanism is starting.

How might this pattern through space and time be explained? Links have been made with plate tectonics. Sinking (tectonic subduction) of the Earth's crust is a way of slowly transferring carbon-rich sedimentary rocks such as shale and limestone deep into the Earth's mantle. Deep beneath the cover of a continent, such ingredients were transformed to give rise both to diamonds and also the inferred propellant for the kimberlite magma: high-pressure carbon dioxide gas, to help it punch through hundreds of kilometres of dense mantle and then crustal rock to the surface. Exactly how this gaseous mixture burrowed all the way up through solid rock remains a mystery. Once it burst out at the surface, diamonds would have rained down out of the sky with kimberlite ash around the erupting volcano.



It seems that before a billion years ago, the mantle was too hot to stew up the particular recipe for kimberlite magma. Plate tectonics itself seems to have developed in its current form just some 3 billion years ago. Before then, the ancient Earth was hotter and seems to have had some other kind of planetary machinery, and other types of volcano. Remnants of these can still be found, among the rocks that remain from those distant, earlier times.

## The Earth's most ancient volcanism

The Earth's oldest rocks occupy the central regions of large continents. Many have been heated and crumpled during later mountain-building episodes, scrambling much of the evidence of how they originally formed. But on a large scale the structures display a distinctive pattern, as in the Barberton region of South Africa and around Pilbara in Australia. Enormous granite-like igneous intrusions are separated by 'greenstone belts'—linear masses of what were basalts and even more iron- and magnesium-rich lavas. The greenstone belt rocks do not seem to represent ancient equivalents of modern ocean crust because their chemistry and physical structure are too different. One hypothesis is that they erupted out of 'heat pipes'—vertical conduits for magma that cut through an early crust that had not yet separated into the independently moving tectonic plates as we have today. If so, this primordial volcanism would be similar to the prolific volcanic activity that now takes place on Jupiter's moon, Io, which also has heat pipes and no plate tectonics. Such an early Earth would have been a quite different planet to the one we know now, with a fundamentally different heat release mechanism.

An interesting kind of lava found in greenstone belts is *komatiite*. Komatiite contains far more magnesium than a basalt: its composition is closer to that of the mantle because when the Earth was hotter, a greater proportion of the mantle melted at plumes. Komatiite emerged at the Earth's surface at something like 1600°C, some 400 degrees hotter than today's basalts and nearly as runny as water (albeit dense and white-hot), so some lava flows were only a centimetre thick, although as with basalt lavas they thickened up in depressions.

Going right back to our planet's origins, perhaps the most spectacular lava of all was a planet-wide lava ocean, perhaps 1000 kilometres deep, immediately after the very early Earth is thought to have collided with a Mars-sized planet, Theia. This gargantuan impact destroyed Theia, which partly combined with Earth, while some of the outflung impact debris coalesced to form our Moon. On Earth, the evidence of quite how this primordial lava ocean solidified into the first crust has now been lost, after 4.5 billion years of continuous change on our singular and dynamic planet. To follow this kind of story, we need to consider volcanism on other planets, as we will do in Chapter 9.

# Chapter 7
# Volcanoes, climate, and the biosphere

Volcanoes keep us alive. Volcanic degassing has provided much of our atmosphere, and it is the alteration of volcanic rock (on land and under the sea) that helps regulate the composition of the atmosphere, notably through the absorption of carbon dioxide in the chemical reactions associated with weathering. Long-term, these processes have kept the Earth habitable for over 4 billion years, even while external influences (such as the heat production of the Sun) have seen significant change.

On shorter timescales, however, volcanism has perturbed both climate and the complement of living organisms on Earth, both locally and globally, nudging biological evolution along. Volcanic outbursts, depending on their nature and scale, can cause global warming or global cooling. In the historical record, even geologically modest eruptions have had—to human eyes—dramatic repercussions. The much larger, less frequent volcanic events that we can deduce from the Earth's geological record have had, we are coming to realize, truly profound effects at a planetary scale.

## Local weather

Volcanoes make their own weather. Water vapour released from the magma into the air is carried high into the atmosphere in eruption columns where it cools and condenses to produce rain.

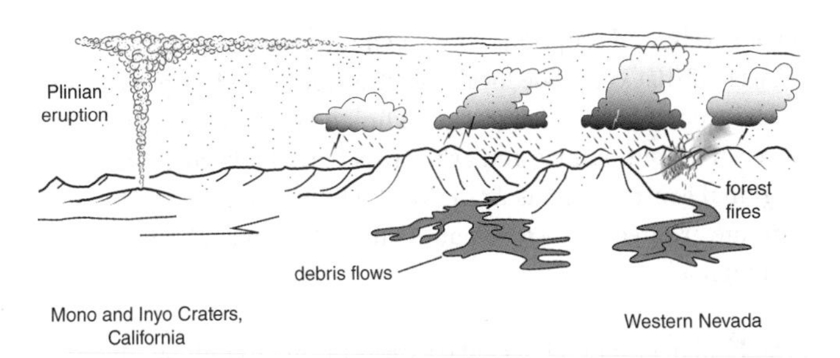


**39. Cartoon of thunderstorms, fires and flood-generated debris flows suggested to have been triggered by the explosive eruption of the Mono and Inyo volcanoes, California.**



This association may be consistent enough to leave a distinct signature in the rock record. In eastern Nevada, repeated Plinian eruptions from the Mono and Inyo craters of eastern California over the last few thousand years left layers of tephra as far as 250 kilometres to the north-east, into western Nevada. There, a curious association has been noted. Each tephra layer is overlain by a thick, boulder-rich deposit: a debris flow unit, the product of an intense flash flood or lahar. In this arid part of the world, the rains occur infrequently, and such flash floods normally only occur once every 500 years or so. To have such an association of tephra and debris flow by chance is highly unlikely, and here it seems plausible that each major eruption also triggered intense rainstorms in mountains many kilometres away (Figure 39). Volcanic eruptions, though, can have an even wider reach.

## Volcanic dimming

Huaynaputina, in the language of the Quechua Indians of what is now southern Peru, means 'young volcano'. Young, perhaps, but its youthful energy made itself felt around the world, despite its unassuming appearance: Huaynaputina is simply a cluster of craters sited within the theatre-shaped collapse depression of a long-extinct volcano. On 19 February 1600 it erupted explosively, and eruptions continued intermittently until the March of that year. The local area was devastated by pyroclastic density currents and lahars, and Plinian eruptions dispersed a layer of fallout pumice and ash across Peru, badly affecting regional centres such as Arequipa, 70 kilometres away. The total volume erupted was equivalent to about 11 cubic kilometres of magma.

The Plinian eruption injected between 15 and 30 million tons of fine ash and sulphur dioxide, as a fine aerosol of acid droplets, high into the stratosphere. Once above the clouds, such an aerosol is not washed out by rain. It can stay at these heights for months to years, where it is dispersed across the globe by winds, until it slowly settles out. Material like this can reflect a fraction of the incoming sunlight—and in doing so it can change the world.



In Russia, it brought crisis to Boris Godunov and the empire that he ruled. Godunov had survived the rule of Ivan the Terrible—and indeed Ivan's attack on him when he had attempted (unsuccessfully) to prevent the tyrant killing his own son. As first a regent, and then a tsar, Godunov had ruled prudently and more or less peacefully. But the particulates flung by Huaynaputina into the stratosphere undermined Godunov's Russia. They induced perhaps the greatest average temperature drop in the northern hemisphere—an estimated 0.8°C—of the last millennium, and deposited more sulphur on the Greenland icecap than Krakatoa was to do nearly three centuries later. Between 1601 and 1603, night-time temperatures in the Russian summer were often below freezing, harvests failed, and winters were bitterly cold. Despite the distribution of money and food by the government, there was famine. About a third of all Russians died, with 127,000 being buried in mass graves in Moscow alone. It was perhaps the key factor in what was called the Time of Troubles in that country, with great social and political unrest. Godunov, after a long illness, died in 1605. His death ushered in the start of the Romanov dynasty.

Explosive volcanic eruptions of sufficient scale to perturb global climate occur once in every several decades. Following Huaynaputina, there came the eruption of Long Island in New Guinea, around 1660, that may have contributed to the Little Ice Age; Tambora in Indonesia in 1815, which certainly caused the 'year without a summer', more famine, some of Byron's bleakest poetry (and, from his challenge to friends in that gloomy year, the writing of *Frankenstein* and of the first vampire novel); Krakatoa, also in Indonesia (and west, rather than east, of Java) in 1883; Santa Maria in Guatemala in 1902; Novarupta (aka the Valley of Ten Thousand Smokes in Alaska) in 1912; El Chichón in Mexico in 1982; and Pinatubo in the Philippines in 1991.

These were enormous eruptions, but yet greater eruptions, with thousands of cubic kilometres of magma erupted in a single event, had even larger climate effects. The major Pleistocene eruptions of Yellowstone were of this scale, as was the eruption at Lake Toba in Sumatra, about 74,000 years ago. The climatic effects of the latter have been estimated to have caused a 3–5°C fall in average northern hemisphere temperature for several years (with higher drops at higher latitudes), and, rather more speculatively, have been implicated in a 'bottleneck' (near-extinction) of humans, an idea that is now contested, and in the triggering of a thousand-year-long cold phase of the Pleistocene Epoch.

While this eruption was undoubtedly colossal, the broader climatic and biological effects, particularly to humans, are less securely based. As regards the human bottleneck, there is genetic evidence of a restriction in the human gene pool far back in history, but the timing of this event from genetic evidence has very wide error bars, and the meagre archaeological evidence suggests little change in the nature of human artefacts from below to above the Toba ash layer in neighbouring Asia. Humans—at least in their ancient hunter-gatherer mode—may have been more adaptable to environmental disruptions than once thought. The proposed triggering of a millennial cold event, too, is inherently difficult to establish because the Pleistocene as a whole was marked by millennial-scale oscillations between warmer and colder climate, particularly in the northern hemisphere. These are the Dansgaard–Oeschger cycles, the cause of which remains mysterious. Whatever mechanism controlled them, a causal link between one such shift and the Toba eruption remains possible, but it is made difficult by the complexity of the Dansgaard–Oeschger transitions, and the difficulty of exactly correlating volcanic and climatic events in the rock record.

## An overdose of basalt

Compared with the great explosive eruptions, the activity of basaltic volcanoes is typically thought of as relatively benign. The frequent lava eruptions of, say, the Hawaiian volcanoes are spectacular and photogenic tourist attractions, but they normally represent little real threat to human life, though at times (as at Kilauea, 2018) destroying property and infrastructure. But there is a question of scale here. Pump enough basalt quickly on to the surface, and with that basalt will come a variety of gases—carbon dioxide, sulphur dioxide, hydrogen chloride, and hydrogen fluoride—and these bring consequences, both for life and for climate.

In 1783, a fissure, some 27 kilometres long, opened at Laki, in south-eastern Iceland, and basalt began to pour out on to the surface. By the time the eruption had stopped, eight months later, some 14.7 cubic kilometres of lava had erupted, covering 580 square kilometres of the island. With the lava came an estimated 120 million tons of sulphur dioxide and 15 million tons of hydrogen fluoride. The air turned foul and 'bitter as seaweed', as described by a remarkable local cleric, Jón Steingrímsson, whose account, *Fires of the Earth*, remains a classic of volcano writing. About a fifth of the Icelandic population died, painfully, many with the characteristic bony overgrowths produced by fluorine poisoning.

The effects went further. The eruption columns ascended into the sky, carrying gas and aerosols across the northern hemisphere and producing a persistent dry fog and blood-red sunsets in that European summer (Benjamin Franklin, indeed, guessed that an Icelandic volcano was to blame). Climate models of this event suggest that it led to cooler summers in North America, but enhanced an already warm summer in western Europe (that then gave way to one of the coldest winters on record). North Africa and India warmed, with a weakening of the monsoon, and the flow of the Nile was reduced markedly.

For Iceland, continually being torn apart in its position astride the mid-Atlantic spreading ridge, such events are part of a pattern. Nearly 800 years earlier, in AD 939, another Icelandic fissure, Eldgjá ('Fire Chasm'), had erupted basalts on a similar scale. The historical records are less detailed, but nevertheless indicate the abandonment of Viking settlements on the island and, farther afield, severe climatic disruption, famine, and political unrest across western Europe, Portugal, the Middle East, and China. Life around a volcanic region, even some distance away, can be precarious.



## A *large* overdose of basalt

Large fissure eruptions of basalt such as Eldgjá and Laki that occur once every few centuries affect climate in a broadly similar way to large silica-rich explosive eruptions. It is the aerosols, notably sulphur dioxide, that in general lead to climate cooling of a few years' duration. The climatic effect of carbon dioxide, though, from even such historically large eruptions, is negligible (in contrast with the much larger carbon dioxide emissions produced by modern human activity).

However, it is becoming increasingly clear that terrestrial volcanism, while exceedingly smooth-running on a planetary scale through the unique (as far as we know) mechanism of plate tectonics, is not *absolutely* regular. Over a timescale of tens to hundreds of millions of years, significant global variations in total global volcanism can occur, as when an ascending mantle plume first impinges on the base of a tectonic plate (Figure 40). The ensuing outburst of volcanism may bring with it sufficient carbon dioxide—trillions of tons—to have long-term effects on global environment and climate (other gases are released also, and hence the effects on climate are complex).

In the warm world of the Cretaceous Period, there were times, each lasting of the order of a million years, when the sea floor became wholly starved of oxygen, more or less worldwide. During these global 'oceanic anoxic events' the sediment that typically lay on the sea floors—white chalky oozes with the remains of burrowing animals—became replaced by black, carbon-rich muds colonized only by anaerobic (non-oxygen-using) microbes, while marine organisms suffered increased levels of extinction. During these times the global sea level was higher, and the climate was even warmer than was the norm. What caused these events?



The first of the Cretaceous oceanic anoxic events took place during the Aptian Age, *c.*120 million years ago. At that time, there was intense volcanic activity in the Pacific Ocean that led to the construction of the Ontong Java Plateau—a 'Large Igneous Province' of basaltic oceanic crust up to 30 kilometres thick (that is, about double the normal thickness) and covering some 2 million square kilometres. The volume of magma going into its construction was immense—some 100 million cubic kilometres—and it has been estimated that the 'extra' gas released by this unusually productive burst of volcanism significantly raised the carbon dioxide levels within the atmosphere, perhaps going as far as doubling them. The ensuing global warming event is thought to have raised sea levels and slowed the (already sluggish) circulation of the Cretaceous oceans, leading to the more rapid exhaustion of oxygen and hence the onset of marine anoxia. In addition, the extra $CO_2$ acidified the oceans, and has been linked with a sudden marked decrease in the building of calcium carbonate skeletons by planktonic organisms.

**40. Thick horizontal lavas eroded to form stepped, trap topography cover the western and eastern fringes of the North Atlantic. They record voluminous basalt eruptions from a mantle plume just before this part of the ocean opened, about 55 million years ago. Ardmeanach, Isle of Mull, Scotland.**



The coincidence in timing of this volcanism and Cretaceous global environmental change is clear—but could this simply be a matter of chance? There is further evidence to pin the blame for the oceanic slowdown on Ontong Java volcanism. The evidence is subtle, and hinges on variations in the proportions of two isotopes of the trace element osmium, which is present in only a few parts per billion in marine sediments. A heavier isotope ($^{188}Os$) is typical of origin from the mantle—and therefore of basaltic volcanism—while a lighter one ($^{187}Os$) is more commonly derived from the weathering of continental rocks. Measuring the proportion of the two in a sample of sediment, therefore, will give an idea of which source—ocean floor basalts or continental rocks—has contributed more to its origin. The sediments formed at the beginning of the Aptian anoxic event show a clear increase in the heavier isotope $^{188}Os$, which is most reasonably explained by release of this isotope into the seawater during the eruption of the Ontong Java basalts.

Links between major eruptive events and long-lived climate perturbations are becoming more apparent. For instance, another global climate event took place 55 million years ago, at the boundary between the Paleocene and Eocene epochs. Chemical evidence from layers of sedimentary rock formed at that time indicate that there was a massive input of isotopically 'light carbon' from the Earth into the oceans and atmosphere, associated with global warming and ocean acidification, in an event that lasted about 100,000 years. Initially, it was thought that this 'light' carbon was mostly derived from non-volcanic sources such as methane vented from hydrocarbon deposits in strata, as such methane has a lighter isotope signature than volcanic $CO_2$, and is an even more powerful greenhouse gas. But more detailed studies suggested that huge amounts of volcanic $CO_2$ led to this pulse of rapid warming, released in a burst of magmatism as part of the North Atlantic Ocean began to split open (Figure 40).



The Cretaceous and Paleocene–Eocene events were not the largest of their kind, though. Most of the major catastrophic shifts in life on Earth seem, too, to be linked to gigantic volcanic outbursts.

## Turning points for life

Life on Earth has been marked by long intervals of benign conditions in which its diversity has increased, over intervals up to tens of millions of years long. However, separating these intervals have been episodes, some abrupt, in which biodiversity tumbled, as many species—and often whole families—of organisms became extinct. These mass extinction events are profound signposts in the geological record, for they separate very different worlds. They have also been profound mysteries for geologists, who have strived to unmask the killing agents. The greatest of these events are commonly referred to as 'the Big Five' mass extinctions: they happened at the end of the Ordovician; in the late Devonian; and at the ends of the Permian, Triassic, and Cretaceous periods.

The most famous of these is that at the end of the Cretaceous, for it involved the demise of the dinosaurs. But the most severe one was at the end of the Permian period, when roughly half of the *families* (involving about 95 per cent of the species) of multicellular animals disappeared forever.

Is there a common thread? When the timing of these biological catastrophes is compared with that of other major events in Earth history, the most consistent link is with major volcanic events. The end-Ordovician mass extinction seems to have been caused by a brief but intense glaciation, and the ultimate cause of the Devonian event, which was associated with marine anoxia, remains mysterious. However, the last three large mass extinctions all happened during the eruption of continental large igneous provinces: the Siberian Traps, the Central Atlantic Magmatic Province, and the Deccan Traps of India, respectively. A catastrophic asteroid impact took place at about the same time as Deccan volcanism, dramatically amplifying the kill factor.

The greatest of the mass extinctions was the one at the boundary between the Permian and Triassic periods, often called the 'Great Dying'. It occurred 251 million years ago, 10 million years after a lesser mass extinction event in the late Permian (also associated in time with the eruption of copious 'flood' basalts at Emeishan in China). It might be said that the Earth was primed for a mass kill, for global climate was hot, and the oceans were sluggish and prone to anoxia. Then came the greatest known incidence of basalt lava pouring out on to a landmass.

The Siberian Traps are almost entirely of basalt that once flowed out over the ancient stable continental landmass of northern Siberia. In the quarter of a billion years since the basalt formed, much has been eroded away, but enough remains to suggest that they originally covered some 5 million square kilometres to a thickness of up to 3 kilometres. The original volume was probably of the order of 3 million cubic kilometres—this is enough to bury the UK landmass to a depth of 12 kilometres.

Obtaining radiometric ages of basalts is technically difficult, but the dates so far obtained cluster closely around 251 million years—at least closely and perhaps exactly coincident (as far as the errors on the dating will allow) with the extinction event. Most of the lava was erupted within less than 2 million years, and perhaps in under a million years. This is a remarkable—perhaps literally unprecedented—rate of outpouring.

What, though, was the kill mechanism? The gases released with the basalts would have been as at Laki, but hundreds of thousands of times greater in scale: carbon and sulphur dioxides, and hydrogen chloride and fluoride. An event of this scale would have released sufficient $CO_2$ to further heat an already hot world, and it is becoming increasingly clear that a major global warming event took place at this time. The direct physiological effect of this is hard to analyse from the fossil evidence that is left to us, but one likely effect of this would be to further stagnate the oceans. One of the most pronounced, and easily readable, signals of the end-Permian crisis is when anoxic conditions affected not just the deep ocean floors, but extended high up on to the continental shelves, suffocating whole communities of shallow-water bottom-living organisms.

Massive release of sulphur dioxide took place, and acidified the oceans. In addition, the fluorine and chlorine emitted with the basalts would have devastated regional vegetation and animal communities, and may have combined into organic compounds that ascended to the stratosphere, to damage the ozone layer. To affect climate, sulphur would have to be carried to more than

10–15 kilometres in the atmosphere. But, being less violently explosive than silica-rich, explosive eruptions, basalts tend to produce lower eruption columns, and to generate tall eruption columns would entail magma discharge (flux) rates significantly greater than has been recorded historically. Ashfall layers that record such a vast atmospheric plume have yet to be discovered associated with these 251-million-year-old lavas.

There is a wealth of potential kill mechanisms. But which were really instrumental in causing mass death, and under which background conditions? To turn plausible *Just So* stories into well-constrained scientific explanations, the devil lies in the detail, and here work is still in progress. Not all large continental flood basalts have caused mass extinction events: the mighty Columbia River basalts, for example, that erupted on to the north American landmass during the mid-Miocene Epoch, seem not to have triggered a global crisis.

Realistic estimates must be made of volcanic gases emitted, and of their likely pathways. With carbon dioxide, for instance, one can start by estimating how much carbon, in total, a body of lava such as the Siberian Traps might release, by making comparisons with modern eruptions: about 18 trillion tons. But not all of that will go into the atmosphere and stay there (were it to do so, the Earth would soon begin to resemble Venus). Some will be removed by photosynthesis, some by dissolving into the oceans, and some by reacting with rocks in chemical weathering. The more time between eruptions, the more time for such natural carbon sequestration to happen—and hence the *rate* of eruption (i.e. the time gaps between successive flows) is crucial. And, as regards the take-up of carbon by oceans, slowdown of the oceans might slow carbon absorption in the short term, but the subsequent burial of non-decayed organic matter will increase sequestration in the long term. This is one detective story, and not the only one in volcanology, in which enquiries will continue for some time yet.

## Climate control on volcanism

Volcanoes can affect climate. But can changes in climate, in turn, affect volcanism? In part this may be by simply modulating the style of an eruption, as when an eruption occurs beneath ice rather than on an open landscape. But there are more subtle mechanisms too. A major physical expression of climate change is the fall and rise in sea level as land ice grows and melts. A coastal or island volcano may be either submerged or emergent, depending on whether the Earth is in a glacial or an interglacial state. It then will be subject to more or less water pressure (which will influence its stability and its tendency to collapse, degas, and erupt) and will be more or less saturated with water (where water encounters magma there is scope for powerful phreatomagmatic eruptions with explosive steam expansion). There is some evidence that coastal volcanoes are more prone to erupt during a change in sea level—whether it is rising or falling—than during sea level stillstands, but more research on this question is needed. Volcanoes are powerful forces of nature, but they can also be sensitive to the world around them.

# Chapter 8
# What have volcanoes done for us?

Volcanoes, like many other natural phenomena, have long been viewed by humans with fear and suspicion. But volcanoes are far less dangerous to people globally than people are to themselves. They are also less dangerous globally than bad weather, cars, cigarettes, floods, earthquakes, crime, disease, pollution, genocide, obesity, alcohol, drugs, and drought. Perhaps it is worth considering for a moment what volcanoes have done for us.

Volcanism has helped make the planet as it is today, by contributing to the chemical composition of the atmosphere and oceans. In the form of hot springs and submarine black-smokers, volcanism may even have facilitated the origin of life on Earth. And thereafter volcanoes have contributed to global evolution by sustaining exotic local ecosystems in isolated locations, such as the Galápagos and Socorro islands, while providing some of the richest, diverse habitats on Earth, such as Pacific coral reefs and atolls, remote island bird colonies, and lush rainforested slopes. Volcanic islands have provided the essential stepping stones for human migration, as in Polynesia and Micronesia and, more recently, as refuelling/provisioning bases for navigators and explorers crossing the Atlantic, Indian, and Pacific oceans.

Volcanoes provide rich, fertile environments for wildlife and agriculture. Many of the world's finest coffees come from volcano flanks in Costa Rica, Colombia, and the highlands of Guatemala, where rich soils, known as andisols, have developed upon thick tephra layers. Some of the best avocados are grown in volcanic regions of Michoacán, Mexico, and macadamia nuts thrive on the nutrient-rich volcanic soils of Hawaii. Tephra-bearing soils are so fertile they can be purchased in bags as fertilizer for use in vegetable gardens. Wines from grapes grown on mineral- and ash-rich volcanic soils have rich flavours and distinctive aromas, such as Sol Lucet Koshu from Mount Fuji; Etna Rossos and Biancos, Willamette wines from grapes grown on basaltic soils in the Dundee Hills of Oregon, and the wines from Assyrtiko grapes grown on Santorini volcano, Greece. The celebrated sweet Passito from the tiny Italian volcanic island of Pantelleria, made from sun-dried Muscat grapes, is said to have been served by the goddess Tanit, on the advice of Venus, to seduce Apollo.



Water supplies in many volcanic regions of the world, for irrigation and consumption, are from wells or tunnels cut into basalt lavas: with their abundant cracks and breccias, they form superb aquifers for groundwater. They are a valuable water source for millions of inhabitants in the drought-prone Horn of Africa, and the isolated volcano Waw an Namus is the site of rare springs in the otherwise parched Sahara. Basalt aquifers are used to irrigate, cultivate, and populate former deserts, transforming, for example, the barren sage plains of southern Idaho into a centre of potato growing. On the ocean islands of Madeira, the Azores and Canary Islands, tunnels driven into basalt lavas tap groundwater to supply networks of irrigation canals that sustain agriculture and tourism. Volcanic groundwater is bottled for drinking on Mauna Loa volcano, Hawaii, on Tenerife, Jeju-Do in South Korea, and in the Chaine du Puy region of France.

Beneath a volcano, the heat causes groundwater to circulate through the rock mass, providing natural hot waters for thermal baths from Japan to Iceland, Río Caliente in Mexico, and tranquil natural saunas on Pantelleria. Such 'hydrothermal systems' are now providing geothermal energy, a green, clean source of electricity generation. Over geological timescales, the circulating hot waters deep within the rock have caused elements such as gold, silver, and copper to be stripped from the magma or the surrounding rocks, and then deposited in concentrated form as ores within mineral veins that we avidly mine to build our technological empires. Volcanoes are also the source of sulphur long used in gunpowder manufacture.

Paleolithic stone axes are commonly made of lava, such as the obsidian of Roche Rossi Lipari, a volcanic island north of Sicily. Stone axes from this site were much prized, traded widely across the Mediterranean. In England, polished stone axes were fashioned out of volcanic tuff from an ancient caldera lake: no less than 566 extraction sites have been discovered around Scafell and Langdale, and the axes were exported across Britain and Ireland more than 4000 years ago. In the Americas, volcanic obsidian, some from lavas and some from fused fine volcanic ashes, was used for arrowheads by Native Americans, and can still be picked up from the ground in many parts of the wild west of the USA and Mexico. The celebrated giant carved heads or Moai of Easter Island were fashioned out of local hydro-magmatic tuff, which is readily carved and not too heavy. The islands themselves are the summits of ocean volcanoes.

Volcanoes have literally formed the building blocks of human civilization. The so-called 'bluestones' of Neolithic Stonehenge are composed of dolerite (diabase) from west Wales, that solidified beneath an ancient line of volcanoes. The Romans and Aztecs both built with volcanic rocks, mostly basalt and ignimbrites. Lava was used in 600 BC to construct the giant carved gate of Tiwanaku in Bolivia. Magnificent cathedrals and public buildings in Naples, Mexico City, Guadalajara, and the silver mining city of San Luis Potosí are constructed from carved blocks of local tuffs and ignimbrites. Ignimbrites are excellent for building as they are easy to quarry and carve, relatively light, and good insulators; they are still quarried in many parts of the world.

Fruit and vegetables are grown in carefully laid out pumice gravel on the many thousands of agricultural terraces of Tenerife. The aggregate lining roads of many volcanic regions, such as Auckland, New Zealand, is made of scoria scooped out of Strombolian scoria cones, so much so that these little volcanoes are rapidly disappearing in developing areas. Troglodytes have long lived in multi-storey caves hewn into soft 2-million-year-old ignimbrites in Cappadocia, Turkey. Volcanoes have provided natural ports for shipping, such as at Thera in Greece (Figure 41), Simpson harbour at Rabaul in Indonesia, Kagoshima bay in Japan, Pozzuoli bay in Italy, Pearl Harbor on Oahu, Lyttleton and Akaroa harbours in New Zealand, and Port of Aden, Yemen.

**41. Volcanic calderas form natural harbours and attractive tourist destinations. Santorini, Greece.**

Historically, volcanoes have provided a vision of hell to scare people into righteous behaviour. Not quite so fearsome are some of the best multi-pitch rock climbs in UK national parks: the classic climbs at Cenotaph Corner, Idwal Slabs, Llwydd and Clogwyn Du'r Arddu in Snowdonia; Gimmer, Dow, Raven, and White Ghyll crags in Cumbria; Rannock Wall near Glencoe; and countless others known to generations of 'crag rats' are of volcanic origin, made of ignimbrite erupted from ancient explosive volcanoes. Culturally, volcanoes have long inspired music (Felix Mendelssohn's *Hebrides Overture* inspired by Fingal's Cave), art (such as J. M. W. Turner's depiction of Vesuvius erupting, and celebrated prints by Hokusai and Hiroshige: Figure 42), novels (Jules Verne's *Journey to the Centre of the Earth* set at Snæfellsjökull volcano; J. R. R. Tolkien's *Lord of the Rings*), travel writing (on Hawaii, by Constance Gordon-Cumming, and Isabella Bird), and poetry (W. J. Turner's *Romance* evoking Chimborazo, Cotopaxi, and Popocatapétl). Many films feature volcanoes (*Jurassic World, The Devil at 4 O'Clock*, and Roberto Rossellini's *Stromboli*), and countless film sets were located on volcanoes, taking advantage of the striking scenery (*Planet of the Apes, One Million Years BC, Close Encounters, Avatar*). Finally, and not least, volcanoes were instrumental in the creation of ice-cream, for it was the combination of ice taken from high on Mount Etna with seasonal fruits growing on the volcano's lower, warmer flanks that led to the first, and arguably still the world's best, Sicilian *gelato*.

So, volcanoes provide humans with abundant materials with which to build and sculpt (Figure 43); clays for pottery; perlite for transformers; pumice to make stone-washed jeans; obsidian for ultra-sharp medical scalpels; metals to mechanize development; rare metals to digitize it; gold to pay for it; drama to inspire; and ice-cream to soothe.

But, volcanoes across the world are also increasingly being exploited for tourism, quarrying, deforestation, and urbanization.

**42.** ***Sugura Street*** **by Utagawa Hiroshige.**

**43. Volcanoes provide materials for arts and crafts. These figures in central Mexico were carved from local ignimbrite.**



Their indigenous flora, fauna, and coral reefs are in rapid decline, their caldera lakes overfished and polluted.

So, on a global perspective, volcanoes have been essential for the evolution of the biosphere, the emergence of humanity, and for the flowering of culture. But more locally they also pose a hazard, one that increases as burgeoning populations encroach ever closer.

## A hazardous living

Volcanoes have been a danger to living organisms throughout geological time (Figure 44) just as they now can threaten nearby human populations. Most inhabitants pass their lives without incident, but as populations expand the numbers at risk from ashfalls, pyroclastic density currents, lahars, and volcanic landslides increase. Yet people are ever drawn to live nearby, partly by the fertile land, but also because the pressures of rising populations and land prices mean that new places to live are sought, often against well-intentioned planning restrictions.

**44. Fossil skeletons of rhinoceros in Nebraska, that were killed and buried 12 million years ago by ash from an explosive super-eruption in the Snake River Plain, nearly 1000 miles away.**



At volcanoes that erupt frequently a cultural legacy of recent eruptions may inform local behaviour and response, but where a volcano has not erupted in the last generation or two, or where the local inhabitants are recent migrants, there may be little appreciation of what devastation may ensue, what warning signs to look out for, and what to do should they appear.

Unlike earthquakes, which can strike without warning, volcanoes commonly provide warning signs before a major outburst, and this significantly improves the potential for mitigating disasters. Locals may detect abnormal animal behaviour, plants dying back, unusual levels of steam or gas emissions, rock falls, or changing temperature or acidity in local springs or streams. Much of this, however, is likely to go unreported or ignored unless a system is put in place within local communities to report, collate, and respond.

Volcanologists may put in place their own systems to monitor changes in volcanic behaviour that might otherwise be overlooked.

Earthquake activity can be monitored by deploying seismometers, swelling and tilting of the ground surface around a volcano may be monitored using repeated laser sightings (LIDAR), GPS systems, and tiltmeters. Spring water chemistry and gas emissions may change prior to an eruption, and can be monitored by repeated water or gas sampling. 'COSPEC' measures gas emissions using light spectra, a technique originally used for assessing air pollution from factories, and webcams can relay visual and thermal information.

Again, a practical means to collect, collate, and synthesize all the results must be set up: the data may be collected on sight, or transmitted directly by radio, or using mobile phone networks to a local volcanic observatory. The deployment of such technology can be expensive and needs maintenance, as unprotected monitoring or transmitting equipment left in the field can be pilfered or damaged.

Satellite monitoring is used to detect changes in temperature, ash plumes, or tiny changes in ground elevation. Measuring the amount and rate that an area bulges, for example, may indicate where magma is accumulating at shallow levels under the ground. The results can be modelled to give an idea of the likely depth and size of a growing magma body. Such 'remote sensing' does not require someone to visit the volcano on a daily basis. Many volcanoes are remote and difficult to access, and it is best to have regular measurements over a long time period, preferably well before an eruption, otherwise one cannot be sure whether particular levels of ground swelling or gas emissions represent a significant departure from 'normal' levels of activity between eruptions, particularly because such 'normal' activity typically fluctuates.

With some 1500 potentially active volcanoes in the world, permanent monitoring of all of them is not achievable, so efforts tend to be prioritized on those volcanoes considered to be particularly dangerous and where there are vulnerable populations and infrastructure (cities, transport links, nuclear power stations). For example, forty-seven volcanoes are continuously monitored in Japan, a relatively wealthy and developed country.

Monitoring will not, in itself, provide all the information needed to mitigate the hazard of a volcano. Even if one successfully forecasts that a volcano is likely to erupt in the near future, monitoring will not determine what type of eruption this will be—will it be a major catastrophe or merely a temporary tourist attraction? Different volcanoes behave in different ways and an individual volcano changes its behaviour from time to time. So it is critically important to involve local geologists with expertise on the behaviour of the individual volcano.



Geologists traditionally infer ancient events from the rock record, and have the maxim: 'the present is the key to the past'. Volcanic geologists, on the other hand, turn this around, and use the (historic and prehistoric) past to inform understanding of present events at a volcano and also what may happen in the future. The best way to do this is to study the rock record—the layers of ash, lava, lahar, and breccia deposits around the volcano, as well as the internal structure of the volcano itself. With meticulous forensic investigation, the history of previous eruptions, their styles, their sizes, and their frequencies, plus any associated landslides and lahars, and the areas affected, can be reconstructed. This provides useful ideas about what may take place should the volcano erupt again, but it takes time and resources, both of which may be in short supply when a volcano is showing signs of unrest. And such investigations are commonly hindered by concealment of the rock record beneath younger strata, vegetation, ice, or water.

As with weather forecasting, methods of monitoring and forecasting volcanic eruptions are improving. Also like the weather, volcanoes are complex systems that can occasionally generate seemingly capricious events.

Successful forecasting will not save lives or livelihoods without well-thought-out planning that must put systems in place to ensure that eruption warnings lead to timely evacuations at short notice when the need arises. Where eruptions are infrequent, local communities and civil authorities may not appreciate the need for detailed plans and regular evacuation drills. Stretched civil authorities often have other crises to deal with, too, from typhoons to terrorists, transport and industrial accidents, tectonic earthquakes or tsunami. This means that disaster awareness and response systems are best instituted well before crises arise. They typically may involve local and regional government, emergency services, including disaster response and the military, together with weather forecasters and the media, as well as local communities.

In 2018 the small village of San Miguel Los Lotes, 13 kilometres downstream from the more or less persistently active Volcán de Fuego stratocone in Guatemala, was tragically overrun by a hot pyroclastic density current of unexpected size. Houses were filled with hot ash and rocks and bulldozed by the current, which shredded even reinforced concrete (Figure 45). A marked change in eruptive activity had been reported by volcanologists, but the authorities did not organize an immediate evacuation. There were few survivors in the village, and more than 400 fatalities—exact numbers are still not known. Passers-by filmed the rapidly unfolding events on mobile phones without perhaps appreciating the immediacy of the danger, and many suffered severe burns. Improved, more robust response systems are now being put in place. A look where the river has now cut through the new deposits shows—just below the old soil—older deposits closely similar to those of the fatal 2018 density currents. Whatever prompted the volcano to generate the density currents evidently had happened before, likely well before the collective memory of the present inhabitants.

Meanwhile, during the same summer, another unexpected change in behaviour was occurring—at Kilauea volcano, Hawaii. The

**45. Reinforced concrete buildings are no protection against block-and-ash flows. The 2018 disaster at San Miguel Los Lotes, near Fuego volcano, Guatemala.**



volcano's underground plumbing system changed, accompanied by dramatic piecemeal subsidence of the caldera at the volcano's summit, and the opening of twenty or so new fissures in Kilauea's East Rift Zone from which emerged new lavas. The lavas covered over 32 square kilometres of the island and engulfed a residential neighbourhood known as Leilani Estates, destroying 700 homes. This event highlights the real impact of even a modest (0.5 km$^3$) sized and relatively non-violent volcanic eruption. After initial emergency evacuations, a more extended recovery period begins, and former residents must decide whether to rebuild, or move to a less hazardous place.

There have been many success stories in hazard mitigation. Thousands of lives were saved by timely evacuation of local inhabitants when Mount Pinatubo erupted in 1991 (Figure 46). Two major US military bases, too, were successfully evacuated. A continued hazard from lahars persisted for years after the eruption, with the successive inundation of adjacent populated

**46. During the 1991 eruption of Mount Pinatubo, Philippines, ashfall and pumice fall layers blanketed slopes, vegetation, and Aeta villages.**



areas in surrounding agricultural plains that had remained relatively unaffected during the eruption itself. Excellent communication with local populations, which included showing on-screen footage of the effects of lahars at community centres, together with effective interaction between the Philippine volcanology institute, PHIVOLCS, and government relief agencies, the media, and the public led to the instigation of evacuation drills. Then, when heavy rainfall triggered detection devices near the volcano, warning signals were sent out and the inhabitants in those particular downstream areas evacuated largely as planned, even at night.

A volcanic eruption cannot be prevented or subdued any more than one can stop a hurricane. This is why evacuation is commonly the best response. Barriers are sometimes engineered to divert lavas and lahars, but with mixed success. Many at Pinatubo were simply overtopped or washed away. But at Sakurajima, near Kagoshima in Japan, lahar channels at one of the world's most frequently active explosive volcanoes are lined with concrete levees and sediment traps, and children wear safety helmets on the way to school, and take temporary refuge from falling ash in roadside concrete shelters.

What perturbed Lake Nyos in Cameroon in 1986 is still not known. It may have been a small landslide or a small eruption that triggered overturning of the lake water, bringing water supersaturated with carbon dioxide to the surface. The carbon dioxide erupted out of the lake, forming a deadly, invisible, ground-hugging plume, some 50 metres thick, to a distance of 25 kilometres, and killing 1746 people and about 3500 livestock. Degassing pipes have since been installed in the lake to try to prevent a recurrence. Meanwhile, investigations at other African lakes have revealed that others, some much larger than Lake Nyos, also contain water supersaturated with carbon dioxide, and periodically undergo degassing events.



In some cases, volcanic eruptions impact human populations and economies far beyond the local area. The little-known Eyjafjallajökull volcano in southern Iceland sent up an ash plume in 2010 that drifted across the British Isles to mainland Europe. The effects of ash on jet aircraft have been known from ash–aircraft 'encounters' for decades: when the abrasive ash enters jet engines it abrades the compressor blades, then melts within the combustion chamber to form solid coatings of welded tuff that block the airflow, causing engine failure. For this reason, for some years, airline pilots were being trained using simulators on what to do should they fly into an ash plume, and transcontinental flights over remote Aleutian and Alaskan volcanoes were cancelled or redirected whenever eruption plumes were reported. However, although the 2010 Eyjafjallajökull eruption was relatively small and the plume rarely exceeded 8 kilometres altitude, it blew persistently across European airports so that nearly all flights had to be cancelled. Seven to ten million passengers were stranded, and costs to airlines were estimated to exceed 2 billion euros. As the eruption continued day after day the costs and disruption escalated, and some airlines might have collapsed had the eruption continued much longer.

This type of volcanic hazard is likely to increase because air travel is increasing worldwide, not because volcanism is increasing. Much has been learnt since the crisis. Tolerances of jet engines to very low concentrations of ash are being investigated, and improved methods of mapping the positions, altitudes, and ash concentrations of atmospheric ash plumes are being developed.

It is worth thinking what would happen in the event of a future giant explosive super-eruption. We need to learn more about how long such explosive eruptions persist, and what their effects are on weather, climate, and biota. The atmosphere may recover after just a few years, but by this time food production would be in crisis. What type of precursory activity might warn us that a super-eruption is imminent, and how much warning would one have? Would it be possible to evacuate entire regions, with several million inhabitants? Where would the people move to? What would the consequences be should the eruption then fail to materialize?

The dangers of volcanoes have—at least to some extent—been scientifically elucidated, and volcanoes are now understood to be a necessary and inevitable part of the mechanism of Earth. They are present on other planets, too.

# Chapter 9
# Volcanoes beyond Earth

Volcanoes are not just an Earthly phenomenon. On the tiny planet inhabited by Antoine de Saint-Exupéry's *Little Prince*, you may recall, there were three volcanoes: two active and one extinct. Each was knee-high to the young protagonist, and they were most useful for heating his breakfast. He swept them out every day, so that they burned gently and regularly—even the extinct one. For, as he said, 'You never know!'

Indeed, you never know. But as we come to observe the Solar System more closely, we are obtaining glimpses—partial, tantalizing—of a range and diversity of volcanism, active and extinct, that extends far beyond Earth. And when we come, one day, to analysing the myriad planets that lie in other star systems, then they, too, look likely to betray an astonishing range and scale of volcanic activity. Some of this will undoubtedly far surpass anything that we see on Earth. But, even those that we can detect in our neighbours are remarkable enough.

## Venus

Venus is, in many ways, the closest parallel to the Earth. It is approximately the same size (it is, in fact, a little smaller, being some 650 km less in diameter than Earth), and has a crust, mantle, and core. But its behaviour is very different, because it lacks water. Its water evaporated and escaped into outer space billions of years ago, as the 'super-greenhouse' of that planet (with over ninety Earth atmospheres' worth of carbon dioxide in its atmosphere) took hold.

Without water, plate tectonics cannot function on a planet, for there is nothing to act as lubricant to ease the friction on descending tectonic plates. And, without plate tectonics, the course of volcanism cannot run so smoothly and evenly. The landscape of Venus was long completely hidden from us by thick clouds of sulphuric acid, and with such a mystery the human mind could run riot, imagining steamy jungles inhabited by exotic Venusian life-forms.

The real landscape, as now charted by radar from the orbiting Project Magellan spacecraft, is lifeless, but no less strange. It is a baking hot desert, at *c.*460°C—hot enough to melt lead. And it is dominated by volcanic landscapes, upon which no rain falls (sulphuric acid raindrops do fall from high in the atmosphere, but they evaporate long before they reach the surface).



There are many more volcanoes on Venus than there are on Earth, and many remain active. In the absence of plate tectonics and the kind of tectonic forces that raise Earth-style mountain belts, and of streams, rivers, and shorelines, it is volcanism and volcanic products that dominate the planet's surface. The radar imaging of the Magellan orbiter has revealed some volcanoes that seem broadly similar to their counterparts on Earth, and some features that seem quite different. Distinctive patterns of radial fractures, termed astra, are thought to represent long-lived fracture patterns that formed, and then became magma-filled, beneath an upwardly ascending Venusian mantle plume. There are also 'pancake domes' (Figure 47): circular, flat-topped, steep-sided volcanoes some kilometres or tens of kilometres across (in truth, they look more akin to crumpets than pancakes). The radar pattern suggests that they have a rough upper surface: they may have formed by viscous

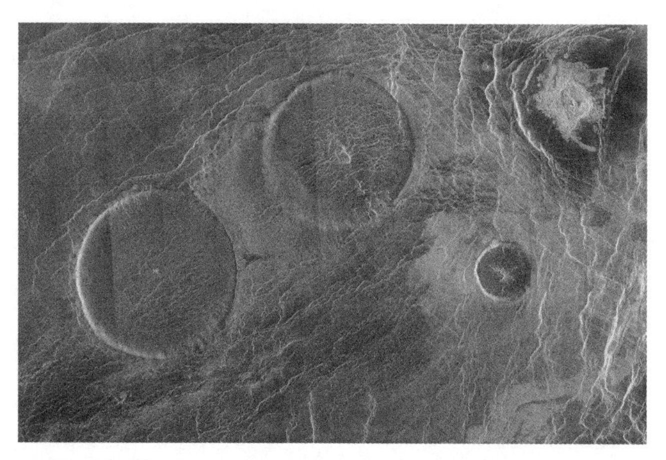

**47. Pancake lava dome volcanoes on Venus, each about 65 km across and 1 km high.**



magma slowly welling to the surface. A variety of these, termed 'tick structures', have strongly scalloped edges, perhaps the result of landslides. There are also enormously long lava channels—over 6000 km long and a kilometre wide in the case of the remarkable *Baltis Vallis*—and some of these include oxbow-like structures and end in lava fans.

This diversity and strangeness of form results from the different conditions in which the volcanic structures are generated. Venus's lack of water most obviously affects the surface but it also affects the Venusian mantle, where melting takes place. On Earth, the small amount of water dissolved in the mantle rock (at least an ocean's worth) greatly influences the production and transport of magma—effectively making the rock melt at lower temperatures and making the resultant magma more mobile. This is especially so where seawater is carried down within the descending ocean plate along subduction zones, partly to be mixed into the mantle, and partly to be erupted back to the atmosphere in the silicic explosive eruptions that characterize these settings.

Plate tectonics does not happen on Venus, and so its patterns of magma generation must be very different from that on Earth. The conditions under which the magma arrives at the surface also differ from Earth. The crushing atmospheric pressure (akin to those at the bottom of the ocean on Earth) and the high ambient temperatures mean that the lava, as it emerges on to the surface, is less likely to lose its volatiles by bubbling out. And, the temperature difference between magma and atmosphere being smaller, the lava will chill and solidify less quickly as it travels. This surely contributes to the length and complexity of some of those surface lava pathways.

We see little fine detail: the radar does not 'see' small features, and the very brief, more detailed glimpses obtained of the Venusian surface by the Soviet *Venera* landers before they were destroyed by the hostile surface environment give little extra information of volcanic process. The surface is too hostile for further spacecraft—or astronauts—to attempt landing in the foreseeable future.

But there are other, grander perspectives on Venusian volcanicity. The whole-planet radar picture shows a curious pattern to the landscape of this planet. Meteorite impact craters are scattered more or less evenly across its surface and seem to represent about half a billion years' worth of collisions. There is no trace of the profusion of the large, cataclysmic impact craters that can be seen on, say, the Moon and on Mars, and that represent the first billion years or so of this Solar System's history, when large asteroids—the debris left over from the planetary construction process—still abounded, and frequently collided with the planets. Something has renewed the surface of Venus.

This renewal process might have simply been more or less steady eruptive activity, gradually filling in the primordial landscape and then piling higher and higher on the surface. But, another, more dramatic process has been invoked—that of 'resurfacing' of the planetary surface every several hundred million years. According to this hypothesis, heat builds up inside the planet, prevented from being slowly and steadily released the way it is on Earth by the marvellous heat release mechanism that is plate tectonics. As the heat build-up continues, magma gathers underground until it is catastrophically released at the surface. Venus is then, in a sense, turned inside out.

It is a spectacle fit for the most spectacular of blockbuster disaster movies. But is it true, or partially true, with Venus undergoing more and less active phases of volcanism on long timescales? The devil lies in the detail of regional crater patterns and large-scale volcanic stratigraphies that are still painstakingly being assembled. Watch this space. Some of the biggest questions on Venusian volcanology remain tantalizingly open.

## The Moon, Mercury, and Mars



Here, we are in the realm of small planets and planetary bodies: those that have lost their internal heat to the point at which magma production has ceased completely, or almost completely. The Moon and Mercury show many resemblances: they lack both water and an atmosphere (for all practical purposes) and their geological activity has now stilled—or almost so. They are only very slowly being abraded by the continual infall of micrometeorites (and occasionally, very locally pulverized by the arrival of a larger meteorite). The volcanic landscapes that we see on them formed billions of years ago, and are now fossilized.

We have the best view of the Moon, our neighbour (and indeed can see some of its geology on a clear night, with the help of a pair of binoculars). The major elements are clear. There are large pale patches—the lunar highlands. These are the remnants of the first magma ocean on that body, the primordial surface that developed on the Moon as it coalesced from a cloud of incandescent vapour flung, it is thought, from the Earth after a giant planetary collision, early in the Earth's history.

The primordial lunar magma ocean cooled, and eventually crystallized. As the crystals formed, the heavier ones—of olivine and pyroxene, rich in iron and magnesium, sank, and the lighter ones—of silica-rich feldspar—rose to the surface, and formed the first lunar crust. How did it behave, this first crust? Did sufficient convection occur in the molten rock below to set up a short-lived form of plate tectonics?

We might never know, for that primordial crust was then pounded by many meteorite impacts in the violent first billion years of the Solar System as it swept itself clear of asteroidal debris by the simple expedient of having this material smash into planets and moons. The lunar highlands now consist of rubble, made of fragments of anorthosite, a rock formed almost entirely of the calcium-rich feldspar anorthite.

Over the next couple of billion years, basaltic magma formed and slowly welled to the surface, to form the dark patches, visible to the naked eye, that are the lunar 'seas' (Figure 48). The oldest rocks recognized of this type date from before a hypothesized 'late heavy bombardment': a flurry of meteorites thought to have raked the inner Solar System some 3.8–3.9 billion years ago, unleashed, it is inferred, by a realignment of Jupiter's orbit. Most of the 'seas' that we can see, though, post-date this inferred event (and hence are well preserved), as lavas that fill low-lying areas and crater basins—with some that likely formed subsequent to major impacts that created pathways between deep-lying magma and the surface. Most of this volcanism ceased more than a billion years ago—but the Moon might not quite be dead, for small patches of disturbed rubble at the surface have been ascribed to the release of deep gas pockets, last gasps from the cooling Moon.

On that crater-scarred, Moon-like planet, Mercury, volcanism is less obvious, because there is less of a contrast between primordial, light-coloured anorthositic crust and subsequent

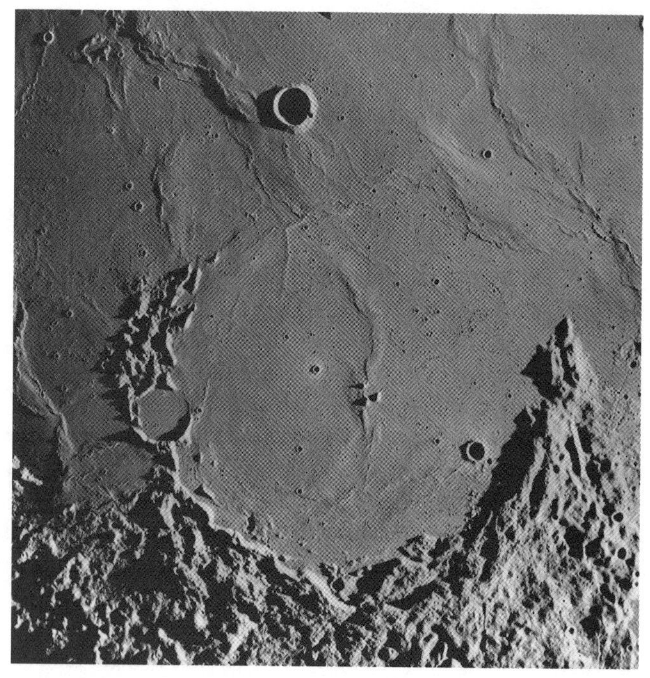

**48. A lunar mare basalt landscape adjoining and overlying the much rougher highlands terrain.**

basaltic lavas. Early fly-bys by the *Mariner* spacecraft, in 1974–5, did little to resolve the presence, or extent, of volcanism on this planet. Smooth plains were imaged between the craters, but there was not enough evidence to say whether these were lava fields or spreads of impact debris. In early 2008, though, the MESSENGER explorer captured more detailed images that provided convincing evidence of volcanism that indeed seems to have been broadly similar to that of the Moon.

Volcanic aspects of Mercury's landscape include irregular craters and depressions of volcanic appearance (impact craters tend to be circular), and low domes akin to shield volcanoes. Lava has flowed around and partly filled impact craters to depths of (at least) hundreds of metres. Perhaps the most arresting feature is the Pantheon Fossae, originally nicknamed 'the Spider': over a hundred graben (paired crustal fractures with depressions between), some exceeding a hundred kilometres long by a couple of kilometres across, radiating outwards from a central crater, like the spokes of a rimless wheel. Its origin is still controversial, but one plausible mechanism is that it formed, rather like the Venusian astra, as the crust was stretched over an upwardly ascending mantle plume.

It is on that other almost-dead rocky planet, Mars, where volcanoes, long ago, grew taller than anywhere else in the Solar System. Perhaps not larger—for, as we have noted, one can interpret the mid-ocean ridge systems of Earth as being colossal, planet-encircling volcanoes.

Mars may have been almost big enough for plate tectonics to start during the first billion years of its history. The Valles Marinaris, that 4000-mile-long scar across the planet, might perhaps represent the failed beginnings of an ocean that never materialized. Overall, though, magmatism on Mars seems focused upon relatively few regions.

One such region is the Tharsis Bulge, a volcanic plateau a little under 4 billion years old. Sitting on top of this are several volcanoes, including the mighty, and ancient, Olympus Mons, some 27 kilometres high, with a 4-kilometre-high, cliff-like scarp around much of its base that raises it above the Martian landscape (Figure 49). Detailed mapping of the topography and cratering history of Olympus Mons has shown that most of its bulk formed by lava outpourings between about 3.7 and 2.5 billion years ago. Major collapses then also gave rise to a surrounding 'aureole' of landslide deposits that, just as with the tall volcanoes of Earth, cover a greater area than does the volcano itself, and that likely also formed its perimeter scarp.

**49. At 22 km high, Olympus Mons on Mars is the tallest volcano in the Solar System. It has a central caldera and a shield-like shape surrounded by steep scarps enhanced by outward-directed landslides.**



Olympus Mons is essentially a tall shield volcano, resembling Hawaii, and likely it formed in a similar way from long-lived magmatism at a 'hot spot' caused by mantle plume. However, in the absence of Earth-like plate tectonics, Olympus Mons did not slowly move across the plume, and so over time it grew some three times higher than Hawaii. After growing for 1.2 billion-years, magmatism has continued at a slower pace. At the summit there is a composite caldera, formed from multiple episodes of collapse related to lava effusions, the latest about 150 million years ago—at the time when dinosaurs walked on Earth. But is Olympus Mons now dead? Probably not, as crater-counts on its slopes suggest that small lava flows on its slopes continued until only a few million years ago. Mars landers have detected 'Marsquakes' too. In its extreme old age, therefore, Mars still contains enough warmth for occasional late volcanism, and—possibly—enough to incubate, somewhere beneath its permafrost crust, any microbial life that might (perhaps) still lurk there.

## Jupiter's hypervolcanic moon

Some of the Solar System's more bizarre volcanism is expressed in its outer regions. Jupiter, Saturn, Uranus, and Neptune, the gas giants, have no solid surface that we can detect, and hence no detectable volcanoes. Orbiting them, however, are a number of moons, which are astonishingly diverse in their behaviour.

These include the Solar System's most volcanically active body—Io, the nearest of Jupiter's moons. Io, in mythology, was a nymph seduced by a wayward Jupiter. Io is only a little bigger than our Moon and hence might be thought, by now, to be dead. But telescope images from Earth showed that it was generally orange-coloured, unlike the greys of its neighbours, and a high heat flow came from its surface. Then, in 1979, the first *Voyager* spacecraft flew past. Its cameras captured a plume of dust and debris silhouetted against this moon's rim—the sign of an ongoing eruption. The second *Voyager* mission, scrutinizing this body in a ten-hour-long take, detected further eruptions and, in 1995, this moon was photographed, along with its neighbours, by the *Galileo* spacecraft, positioned in orbit around Jupiter. Its cameras showed a volcanically active world (Figure 50).

**50. A volcanic eruption captured on Io, a moon of Jupiter.**

Io is akin to a rocky planet in composition, being a mixture of iron and silicates, and has many volcanoes scattered about its small frame. Eruptions are frequent and spectacular. With no atmosphere, and weak gravity, the volcanic plumes reach to some hundreds of kilometres above the surface, and the erupted material coats the entire surface of the planet. Io has little or no water or carbon dioxide: their place is taken by sulphur dioxide and sulphur, frozen deposits of which coat much of this moon's surface. Indeed, it was first thought, from the *Voyager* data, that the eruptions were chiefly of sulphur compounds. However the *Galileo* data showed the erupted material to be primarily of silicate (mafic and ultramafic) composition, their temperatures reaching 1600°C at the surface, hotter than lavas at the Earth's surface. The continuing volcanism has resurfaced Io, and obliterated virtually every impact crater on it.



The heat source in this case is not internal radioactivity. It is the tides raised as Io is tugged between the grip of Jupiter and the gravitational effects of the neighbouring larger moons, Europa and Ganymede. Io is kneaded like a ball of dough, its surface rising and falling by up to 100 m, and this continual distortion is translated into heat and hence into magma generation.

The captivity of Io, too, is only temporary. Ultimately (in hundreds of millions of years), it will likely escape from the gravitational grasp of its neighbouring moons. When this happens, the extreme tidal squeezing will end, and so will Io's spectacular volcanism.

## Ice volcanoes of the Solar System

The other moons that orbit the giant gas planets of our Solar System are less severely affected by tidal friction, and do not reach such high internal temperatures. They are frigid worlds surfaced by ice—water ice, methane ice, nitrogen ice, or mixtures of these—and their *cryovolcanism* is typically expressed where liquid or gas escapes from beneath the frozen crust.

Each of these moons is different in detail. Europa, Io's close neighbour, for instance, is a world with a bright white crust of water ice, overlying a continuous ocean perhaps 100 kilometres deep, beneath which is a rocky core. The ice crust, perhaps tens of kilometres thick, is 'new', with few or no meteorite impact craters. It is pervasively fractured and grooved, some parts appearing to represent shattered ice floes held within newly formed ice, while thin zones with parallel stripes—'triple bands'—cut across these. These 'triple bands' have been compared with mid-ocean ridges, with successive phases of extensional fracturing allowing water, or brine, or perhaps ductile ice, to ascend and then to freeze onto the fracture margins. Such phenomena are purely cryovolcanic, but Europa might also possess some true silicate volcanism. The tidal stresses it experiences, while not approaching Io's, may be sufficient for such volcanism to occur on the rocky floor of this moon's ocean, deep below the ice crust. Some surface features on Europa may reflect such a process: the more chaotic, floe-like terrains on this moon might, perhaps, have been produced by rising plumes of volcanically heated water.

Other moons seem to possess comparable features. Titan, Saturn's largest moon, is shrouded by a haze of hydrocarbons in its thick atmosphere of mainly nitrogen with a little methane. This haze was penetrated by radar from the *Cassini* spacecraft in 2004, and then, physically, by the *Huygens* probe that settled on the moon's surface early in 2005. Titan, like Europa, has a 'young' icy shell above a deep ocean, beneath which is a rocky core. Nevertheless, this young surface has substantial topography, which is slowly being eroded by methane rain. This topography includes a dome, about 180 kilometres in diameter, that may be the ice equivalent of a large shield volcano.

Activity has been seen on some of these distant icy moons, caught by the cameras on the *Voyager* and *Cassini* spacecraft. Geysers of water, up to hundreds of kilometres high, erupt from Enceladus, one of Saturn's small moons, just 500 km across. They originate from fracture systems in its icy shell, and consist of water together with ammonia, the latter acting as 'antifreeze' allowing the water to remain liquid down to −100°C. Nevertheless, Enceladus is puzzling, for its surface is mostly far colder than that, at −200°C. Is there a reservoir of warmer, well-insulated deep water supplying the geysers, in a scenario dubbed 'Cold Faithful', after the celebrated geyser in Yellowstone National Park? Or is this moon entirely frozen, the water being simply locally generated along fractures by tectonic activity (the 'Frozen Faithful' model)? The former idea is favoured at present, again with tidal stresses being the energy source.

How far out does volcanism extend, in this Solar System of ours? Cryovolcanic plumes have been observed on Triton, a moon of Neptune, as the *Voyager* spacecraft passed by. More recently, in 2015, NASA's *Dawn* mission flew past the dwarf planet Ceres, the largest object in the asteroid belt between Mars and Jupiter, a body nearly 1000 kilometres across, pockmarked with meteorite impact craters. Among these, Ceres has a single steep-sided mountain some 5 kilometres high, named Ahuna Mons, that is very likely an ice volcano, where the 'magma' consists of water ice with mud, softened by included salts (Figure 51). The same year, the more distant dwarf planet Pluto was imaged by the NASA *New Horizons* space probe as it sped by, and its surprisingly varied topography also seemed to include cryovolcanoes.

There are hints that such phenomena might extend much farther out. Beyond Pluto lies the Kuiper Belt of orbiting bodies, and one of these, some 650 km across, has been christened Quaoar. At about 50°C above absolute zero, Quaoar should be made of amorphous rather than crystalline ice, because the ordered state of ice crystals needs sufficient energy to form. Yet crystalline ice has been detected on this body—and perhaps ammonia ice too. It has been suggested that ice composed of a mixture of water and ammonia crystallized deep inside Quaoar, from heat supplied by radioactive decay in its rocky core, and that this then rose along

**51. A volcano built of ice—a cryovolcano—on the dwarf planet Ceres.**



fractures to the surface, in a manner akin to basaltic volcanism on Earth. This is, for now, speculation—albeit *reasoned* speculation. Nevertheless, it demonstrates the truly cosmic range of the phenomenon of volcanism—and suggests that soft ice may be at least as common a type of magma as incandescent rock silicates.

## Beyond the Solar System

In 1992, the first planet to orbit a star other than our Sun was detected. In 1995, another was found. A few years later, the trickle of discoveries of these exoplanets became a flood, helped by specially constructed exoplanet-hunting satellites being launched into orbit. Now, more than 4000 exoplanets are known, many within distant planetary systems that are very unlike our own. They include 'hot Jupiters' with freakishly looping orbits taking them alarmingly close to their parent stars, and 'super-Earths', rocky planets several times bigger than our own. They are so distant that we cannot yet see any surface details, but speculation on their potential volcanism has already begun, as volcanoes are certain to be a common feature of the cosmos, given their near-universal function as a planetary heat release mechanism.

Indeed, as we put the finishing touches to this book, the first plausible evidence of an exomoon has been found from 550 light years away—betrayed, it is inferred, by volcanic plumes of sodium from Io-like volcanicity on its surface. Such distant volcanoes may have a wider significance, too. Volcanically emitted hydrogen on a planet would increase the chances of finding life there, for instance. As the next generations of satellites and telescopes are trained on those distant regions, they will be looking harder for such clues.

And these distant volcanic landscapes, when they are detected, will likely spring many extraordinary surprises. But there is still much to discover, too, about volcanoes much closer to home. These awe-inspiring phenomena, both fearsome and life-giving, will continue to fascinate their human observers long into the future.

# A short glossary

**a'a:** A very rough type of lava with jagged, vesicular rubble and sharp rock spines, formed when a lava flow tears itself apart. Commonly basaltic (Figure 26).

**andesite:** Lava with a composition intermediate between that of rhyolite and basalt. It contains more silica and aluminium than a basalt but less than a rhyolite. It commonly forms block lavas, as in the Andes.

**arc:** A row of volcanoes above a subduction zone, near the margin of a tectonic plate (Figure 1).

**ash:** Dust to sand-sized volcanic particles erupted from a volcano. It commonly comprises explosively fragmented and chilled magma, and bits of the volcano.

**basalt:** Dark lava rich in iron and magnesium and with less silica and aluminium than an andesite or rhyolite.

**block lava:** Lava that is topped by faceted, angular blocks formed by brittle breakage during highly viscous flow. Commonly made of andesite or rhyolite (Figure 27).

**breccia:** A rock made of angular blocks, e.g. scree or parts of a block lava.

**caldera:** A large (>1 km diameter) topographic depression formed by near-vertical collapse of the ground, where supporting magma in an underground reservoir erupts or drains away (Figure 41). Larger than a crater.

**conduit:** A natural pathway or crack through the ground along which volcanic material travels, for example to the surface during a volcanic eruption.

**crater:** A flared pit in the ground blasted out by a volcanic explosion (Figure 16). Not to be confused with a caldera, which is much larger.

**debris-avalanche:** A catastrophic type of giant landslide that forms when an entire sector of a volcano suddenly collapses downslope in the form of internally fragmented blocks.

**dyke:** A thin, screen-like body of igneous rock formed by the propagation of a steep, cross-cutting crack with injection of magma into the Earth's crust.

**entrail pahoehoe:** A type of lava that resembles glistening entrails, formed from the protrusion of thin, runny lava tubes. As seen on Hawaii (Figure 24).

**fault:** A crack in rock caused by breakage and slippage, often causing an earthquake.

**ignimbrite:** The loose or solid deposit of a ground-hugging, ash- and pumice-bearing pyroclastic density current. May form vast landscape-burying sheets.

**intrusion:** A rock body formed underground by the cooling and solidification of magma.

**lahar:** A slurry of sediment and water that flows swiftly down valleys from a volcano. May vary in consistency from very dirty water to dense sludge.

**lava:** The hot material in a lava flow, or the igneous rock formed when the flow stops and solidifies.

**lava dome:** A stubby hill of lava that was too viscous to flow far from the vent. Cracked, and commonly surrounding by accumulated lava blocks.

**lava flow:** Flow of hot viscous rock across the planet surface. The flowing lava may be partly liquid with crystals and bubbles.

**maar:** A volcanic explosion crater in bedrock formed during a violent Taalian eruption. May fill with water to form a small circular lake.

**magma:** Hot, fluid or semi-fluid rock formed locally within or below the Earth's crust. It cools and solidifies to form an igneous rock. Magma commonly contains suspended crystals and gases.

**mantle:** The thick layer of silicate rock within the planet beneath the outer crust, and above a central metal-rich core. The mantle makes up 84% of the volume of the Earth.

**phreatic eruption:** Explosive eruption caused by the violent expansion of underground steam. Phreatic deposits (ash and blocks) are formed of blasted bedrock aquifer.

**phreatoplinian:** The most violent type of large explosive eruption in which fragmenting magma foam interacts with shallow lake or sea water, producing widely dispersed fine ash deposits (Figure 20). Commonly andesitic or rhyolitic.

**plate:** Relatively rigid upper skin of the Earth, made up of the 'lithosphere', which is the crust and the uppermost part of the mantle (see Figure 1). A tectonic plate may slowly move in a direction different to that of the surrounding plates.

**Plinian:** A sustained, highly explosive style of eruption that carries ash and pumice fragments high into the stratosphere, principally by thermal convection (Figure 6).

**pumice:** Light microfoam of volcanic glass, which typically floats on water.

**rhyolite:** Silica- and alumina-rich volcanic rock. Typically forms block lavas, domes and spines, and may be glassy (obsidian), microfoam (pumice), or micro-crystalline (felsite).

**scoria:** A volcanic rock full of visible vesicles, which formed as bubbles when the rock was largely molten. Doesn't float in water.

**scree:** An accumulation of rocks that tumbled from a steep rock face.

**sill:** A thin, screen-like body of igneous rock formed by the propagation and filling of a near-horizontal crack with magma, now solidified.

**Surtseyan:** A style of explosive eruption in which erupting hot magma interacts vigorously with water and loose watery substrate within a lake or shallow sea, ejecting pulsatory slurries of wet tephra and steam (Figure 18). Commonly basaltic.

**Taalian:** A style of violent phreatomagmatic eruption in which hot magma interacts explosively with a groundwater aquifer, commonly forming a maar and surrounding tuff ring.

**tephra:** Loose volcanic ash, scoria, or pumice.

**tuff:** Hardened volcanic ash.

**tuff ring:** A low rampart or ring-shaped volcano made of tuff around a volcanic crater formed during Surtseyan or Taalian eruptions (Figure 17).

**tuya:** A flat-topped, lava-capped volcano that grew through an ice sheet (Figure 22).

**vog:** Smog or haze containing volcanic gas and aerosols. Drifts downwind from a volcano and may cause respiratory problems, as on Hawaii.

**vulcanian:** A style of eruption that characteristically produces a series of short energetic explosions sending columns of ash ~10 km into the atmosphere. Common on andesite volcanoes.

# Further reading

Francis, P. W. (1976) *Volcanoes*. 368 pp. Penguin paperback. An entertaining read—now out of print.

Francis, Peter W., and Oppenheimer, C. (2004) *Volcanoes*. 2nd edn. 536 pp. Oxford University Press.

Lockwood, J. P., and Hazlett, R. W. (2010) *Volcanoes: global perspectives*. 541 pp. Wiley-Blackwell.

Lopes, Rosaly C., and Carroll, Michael W. (2008) *Alien volcanoes*. 176 pp. Johns Hopkins University Press.

Pyle, David M. (2017) *Volcanoes: encounters through the ages*. 208 pp. Bodleian Library, UK.

Schmincke, Hans-Ulrich (2012) *Volcanism*. 340 pp. Springer.

Upton, Brian (2015) *Ancient volcanoes and the making of Scotland*. 2nd edn. Dunedin Academic Press.

Winchester, Simon (2004). *Krakatau: the day the world exploded: August 23, 1883*. HarperCollins paperback.

# Index

For the benefit of digital users, indexed terms that span two pages (e.g., 52–53) may, on occasion, appear on only one of those pages.

## A



## B



## C

## D



## E



## F



## G

# H



# I



# J



# K



# L

## M



## N



## O



## P

## Q



## R



## S

## T



## U



## V



## W



## Y

# GALAXIES

## A Very Short Introduction

John Gribbin

Galaxies are the building blocks of the Universe: standing like islands in space, each is made up of many hundreds of millions of stars in which the chemical elements are made, around which planets form, and where on at least one of those planets intelligent life has emerged. In this *Very Short Introduction*, renowned science writer John Gribbin describes the extraordinary things that astronomers are learning about galaxies, and explains how this can shed light on the origins and structure of the Universe.

www.oup.com/vsi

# GEOGRAPHY

## A Very Short Introduction

John A. Matthews & David T. Herbert

Modern Geography has come a long way from its historical roots in exploring foreign lands, and simply mapping and naming the regions of the world. Spanning both physical and human Geography, the discipline today is unique as a subject which can bridge the divide between the sciences and the humanities, and between the environment and our society. Using wide-ranging examples from global warming and oil, to urbanization and ethnicity, this *Very Short Introduction* paints a broad picture of the current state of Geography, its subject matter, concepts and methods, and its strengths and controversies. The book's conclusion is no less than a manifesto for Geography' future.

> 'Matthews and Herbert's book is written- as befits the VSI series- in an accessible prose style and is peppered with attractive and understandable images, graphs and tables.'
>
> Geographical.

www.oup.com/vsi

# LANDSCAPES AND GEOMORPHOLOGY

## A Very Short Introduction

Andrew Goudie & Heather Viles

Landscapes are all around us, but most of us know very little about how they have developed, what goes on in them, and how they react to changing climates, tectonics and human activities. Examining what landscape is, and how we use a range of ideas and techniques to study it, Andrew Goudie and Heather Viles demonstrate how geomorphologists have built on classic methods pioneered by some great 19th century scientists to examine our Earth. Using examples from around the world, including New Zealand, the Tibetan Plateau, and the deserts of the Middle East, they examine some of the key controls on landscape today such as tectonics and climate, as well as humans and the living world.

www.oup.com/vsi

# PLANETS

## A Very Short Introduction

David A. Rothery

This *Very Short Introduction* looks deep into space and describes the worlds that make up our Solar System: terrestrial planets, giant planets, dwarf planets and various other objects such as satellites (moons), asteroids and Trans-Neptunian objects. It considers how our knowledge has advanced over the centuries, and how it has expanded at a growing rate in recent years. David A. Rothery gives an overview of the origin, nature, and evolution of our Solar System, including the controversial issues of what qualifies as a planet, and what conditions are required for a planetary body to be habitable by life. He looks at rocky planets and the Moon, giant planets and their satellites, and how the surfaces have been sculpted by geology, weather, and impacts.

"The writing style is exceptionally clear and pricise"

Astronomy Now

www.oup.com/vsi